Gauging the Solar System

Measuring Solar System Values for Yourself

Second Edition

GW00456670

Gauging the Solar System

Measuring Solar System Values for Yourself

Second Edition

E. Roger Cowley

Blue Eagle Press

ISBN 9798706607845

To the memory of my wife Sandra

About the author

Dr. Cowley is an emeritus professor of physics and chemistry from Rutgers, the State University of New Jersey. During his academic career he published more than eighty research papers and book chapters on condensed matter and statistical physics and also received multiple awards for his teaching. He has a long-standing interest in amateur astronomy. He is a member of the West Jersey Astronomical Society (www.wasociety.us) and served for several years as its president.

Cover illustrations:
Front cover – Old Moon, Venus and Mercury, February 6, 2016.
Back cover – Observing the transit of Mercury, May 9, 2016

Contents

Appendices

Introduction

I undertook the projects described in this book because of a deeply held belief of mine that one of the best ways to understand a set of scientific ideas is to measure as many as possible of the values involved for yourself. I am supported in this by a well-known physicist:

"I often say that when you can measure what you are speaking about, and express it in numbers, you know something about it; but when you cannot measure it, when you cannot express it in numbers, your knowledge is of a meager and unsatisfactory kind..." Lord Kelvin (1883)

Of course, a measurement of the numbers involved in an area such as atomic or nuclear physics requires equipment that is not accessible to most of us, but that is not the case for much of astronomy. Many values can be measured with little or no equipment beyond patience, a reasonable pair of eyes (mine are not great) and maybe a digital camera and a pair of binoculars. The distinction between synodic and sidereal periods, or the exact meaning of greatest eastern or western elongation becomes much clearer when you have followed the Moon or a planet through its orbit to measure the quantity involved. Do you really understand what a draconic month is? I didn't until I measured it.

The title of this book might have been "Astronomical Projects for Dirty Skies". I live in Cherry Hill, New Jersey, under the umbrella of light pollution surrounding Philadelphia. In my back yard, on a very good night, the limiting visual magnitude of the stars I can see is about 3.5, but on a night when the transparency is poor I need binoculars to see Polaris (magnitude 2.2) and, sadly, it is getting worse. I do sometimes get out into the Pine Barrens of New Jersey to look at faint fuzzies, but there is a definite advantage in being able to do some of my observing nearer to home. Also, I like to give myself fairly substantial projects that direct my observing for months or even years at a time, and many of them are conveniently done near home. I claim no originality for any of these projects. I have tried to include references to other published descriptions or relevant web pages where I know about them.

My astronomical interests lean in two directions that color my choice of projects. Several of them are attempts at the recreation of astronomical history. I found that most of the projects were to find quantitative values of the parameters of the solar system, and this led to the actual title of the book. I have tried to put myself in Copernicus's or Kepler's shoes to understand the working of the solar system. The emphasis on what I call the clockwork

1

of the solar system rather than the physical appearance of the planets may also be a result of my poor viewing conditions. A consequence of this emphasis is that only a few of the projects involve the use of a telescope. I have been surprised to find how many things I could not only see but measure quantitatively without a telescope. Exceptions were the measurements of the angular size of the planets, the observation of Jupiter's moons, and building a Galilean telescope to look at the things that Galileo looked at. In that last case the telescope *was* the project. Most of the observations can be done naked eye or with a pair of binoculars, but in the last few years I have relied more and more on taking digital photographs of the sky. They create a permanent record, and digital images are very easily measured to allow the calculation of angles between objects in the sky.

I do encourage everyone to explore this method of making measurements. I take star-field photographs that include the particular planet or other object that I am measuring and at least three other identifiable stars. I measure their digital coordinates in the photograph and enter them into a spreadsheet that I wrote. This immediately gives the right ascension and declination of the object. As the quality of viewing in my backyard deteriorates the photograph is quite often of what looks like a blank sky to the naked eye. The spreadsheet can be downloaded from my web page at www.GaugingtheSolarSystem.com.

Of course, this bit of technology was not available to astronomers of centuries before this one, but with this exception I have tried to keep the equipment I used as simple as possible, and, in many cases, home-made. In order to do this, I have relied on measurements made over a long period of time. I have measured the length of a sidereal day with an uncertainty of 0.01 seconds, and the average length of a sidereal month with an uncertainty of about 36 seconds. To do these, the first set of measurements was spread over five summers, and the second over 34 years!

The second emphasis I have is on astronomical computing. I belong to a generation that learned to program on room-sized computers with one-off names such as, in my case, EDSAC II, and later worked with the early 8-bit home computers that came with some version of Basic in the ROM and not much else. While I have used many computer languages in my physics research, I still like to use Basic for the, mostly short, programs that I write to assist me in these projects. I treasure the book "Astronomy with your personal computer" by Peter Duffett-Smith, which contains some forty Basic subroutines for astronomical applications. If you are happier with C++ or Java, you can translate them, but there is a very pleasant, nicely old-fashioned, version of Basic still commercially available (in 2020). It is

TrueBasic, www.truebasic.com, produced by the original inventors of the Basic language, John G. Kemeny and Thomas E. Kurtz. If you shun the use of other people's subroutines, the algorithms are all in "Astronomical Algorithms" by Jean Meeus. Many of the calculations can be done on a spreadsheet instead, and if you prefer a spreadsheet to a computer program, the book "Practical Astronomy with your Calculator or Spreadsheet", by Peter Duffett-Smith and Jonathan Zwart, comes with a set of high quality subroutines for Excel. Probably the most computer-dependent part of the analysis that I do is to extract the right ascensions and declinations of the planets and of the Moon from star-field photographs taken with a DSLR camera. I do the calculations using either a TrueBasic computer program or an Excel spreadsheet. They are discussed in chapter 11.

I must warn you that I have used quite a lot of mathematics where I thought it was the best way to say something. The math was essential to me when I developed the analysis, so it seemed proper to include it. In most cases I have put the mathematical parts as separate appendices at the ends of the chapters, so that you can choose to leave it out, and there are several appendices at the end of the book on more general ideas. I do encourage you to dip into some of the math. There is an enormous satisfaction in seeing that the numbers confirm a picture. For example, you might measure separately the synodic and sidereal periods of a planet, and then compare them to the formula that relates them to the Earth's sidereal year. I calculated the size a pinhole had to be to resolve sunspots on the Sun and then observed them with a lensless system. I found expressions for the field of view of a Galilean telescope under different conditions, and then tested the formulas in the field. We get plenty of cloudy evenings when "doing" astronomy has to involve something other than looking through a telescope. Working out numbers, perhaps analyzing a photograph to find the right ascension and declination of a planet, adds to your understanding. I realize that the extra step of rearranging an algebraic equation to solve for the quantity you want to know does take some practice, but it is not so difficult. I can't really say that it is not rocket science, but it is only a small part of rocket science! A bit of applied math that I have made heavy use of is the Least Squares method of fitting parameters to a formula. I have included a separate appendix, appendix D, just on this topic. There is also a philosophical question involved in the almost universal use of mathematics in science. Why does mathematics work as well as it does in describing the Universe? If you are interested in that aspect, take a look at appendix E: "The line goes through the points".

I have been doing this sort of astronomy for a long time. Some of the observations and measurement described here go back over thirty years.

Why did I decide to write about them now? Probably the biggest push was that I retired from my position as a physics professor and finally had the time to collect them all together and realize that they made several binders-full of observations.

Some of my projects are straightforward or short. Two that I have enjoyed are observing New Moons and Old Moons, and determining my latitude and longitude. These are described in the first chapters of parts I and II respectively. Part III begins with an almost ridiculously simple experiment to estimate the surface temperature of the Sun. Part IV begins with looking at the spectrum of the Sun with very simple apparatus. More elaborate projects in part IV are the making of pinhole images of sunspots, building and using the Galilean telescope, and plotting the magnitudes of four variable stars. But the heart of the book, and what I have called "gauging the Solar System", is a lot of small projects collected in parts I, II and III. I have grouped the projects mostly by the type of variable I was measuring, rather than the chronological order in which I made the measurements. As a result, I sometimes refer you forwards to observations made earlier but described later. Please bear with me on this.

I strongly recommend a publication, the "Observer's Handbook" of the Astronomical Society of Canada. I buy a copy every year. As well as an almanac for the astronomical events of the year, it contains a collection of values of physical parameters, such as the periods and radii of orbits for the planets, and various units of time such as the sidereal day and year. When I look at that data, I ask how many of those numbers I can check for myself. There are others who ask themselves the same sort of question as I did. Three books that cover some of the same ground are:

"Do-it-yourself Astronomy" by Sydney G. Brewer, Edinburgh University Press, 1988.

"Measure Solar System Objects and Their Movements for Yourself" by John D. Clark, Springer, 2009.

"Astronomical Discoveries You Can Make Too!" by Robert K. Buchheim, Springer, 2015.

(The first two of these are aimed at the individual do-it-yourselfer, as is mine. The third is more slanted to a laboratory course in connection with a formal course in astronomy. It is more wide ranging than the others and does contain a lot of detailed instructions and ideas.)

Over the last thirty years I have invented or read of methods of measuring quite a few of the numbers characterizing the planetary orbits, and I collect them here under the title of gauging the solar system. Of course, most of my values are not as precise as the values in the Observer's Handbook, although I have several times been surprised to find how precise they could be. I never expected to match the handbook, and I don't mind. I do these observations for two reasons. One is that I am continually learning new bits of astronomy and the other is that it is fun. Occasionally, in connection with some method, I ask myself if I am cheating. That is, am I really building into my observations some data from outside that already contain some or all of the result. I am sure that the answer is frequently "yes" in some people's judgment. For example, when I found the radius of the Earth, I used a distance between two points that I had measured from a map. I did make a measurement for myself but it was not really accurate enough. We are so surrounded with information nowadays that it is difficult to avoid. The daily newspapers carry the times of rising and setting of both the Sun and the Moon, and sometimes information about which planets are in the sky. The astronomy magazines give much more detail, and then there are the computer programs that will give you more or less reliable values for almost any astronomical number. Initially I relied a lot on these sources but, as the range of numbers I had measured increased, I was able to predict many events for myself with adequate accuracy. I think the decision of whether a particular approach is cheating is a completely personal one. I do this for the fun of it, and as long as it is fun then I am not cheating.

Here are my current best values for some solar system parameters:

A Synodic Month (new moon to new moon) is 29.530 ± 0.001 days.
A Sidereal Month is 27.3214 ± 0.0004 days.
A Draconic Month is 27.212 ± 0.002 days.
An Anomalistic Month is 27.555 ± 0.001 days.
A Sidereal Day is 23 hours 56 mins 4.10 secs ± 0.01 secs.
From this, a Sidereal Year is 365.26 ± 0.01 days.
A Tropical Year is 365.245 ± 0.01 days.
The ecliptic is inclined at an angle of $23° 26' \pm 5'$ to the equator.
The eccentricity of the Earth's orbit is 0.016 ± 0.002.
The Synodic Period of Mercury is 115.9 ± 0.3 days.
The Synodic Period of Venus is 584.0 ± 0.25 days.
The Synodic Period of Mars is 776 ± 3 days.
The Synodic Period of Jupiter is 398.7 ± 0.6 days.
The Synodic Period of Saturn is 377.8 ± 0.2 days.
The Synodic Period of Uranus is 369.5 ± 0.4 days.
The Sidereal Period of Mercury is 88.0 ± 0.1 days, or 0.2409 ± 0.0003 yrs.

The Sidereal Period of Venus is 224.68 ± 0.03 days, or 0.61514 ± 0.00008 yrs.

The Sidereal Period of Mars is 1.89 ± 0.01 yrs.

The Sidereal Period of Jupiter is 11.91 ± 0.06 yrs.

The Sidereal Period of Saturn is 30.6 ± 1.0 yrs.

The Sidereal Period of Uranus is 87.6 ± 1.0 yrs.

The radius of the Earth is 4200 ± 300 miles or 6750 ± 500 km.

The mean distance to the Moon is 60 ± 3 Earth radii, or $250,000 \pm 12,000$ miles.

The diameter of the Moon is 0.27 ± 0.02 times the diameter of the Earth, or 2300 miles.

The mean distance of Mercury from the Sun is 0.38 ± 0.01 a.u. (astronomical units).

The mean distance of Venus from the Sun is 0.723 ± 0.004 a.u.

The mean distance of Mars from the Sun is 1.524 ± 0.004 a.u.

The mean distance of Jupiter from the Sun is 5.21 ± 0.01 a.u.

The mean distance of Saturn from the Sun is 9.72 ± 0.16 a.u.

The mean distance of Uranus from the Sun is 19.5 ± 0.1 a.u.

The diameter of the Sun is $(9.28 \pm 0.05) \times 10^{-3}$ a.u.

The diameter of Mercury is 0.0036 ± 0.0004 times the diameter of the Sun.

The diameter of Venus is 0.0088 ± 0.0002 times the diameter of the Sun.

The diameter of Mars is 0.0051 ± 0.0008 times the diameter of the Sun.

The diameter of Jupiter is 0.104 ± 0.002 times the diameter of the Sun.

The diameter of Saturn is 0.088 ± 0.006 times the diameter of the Sun.

The mass of the Earth is $(6.7 \pm 0.8) \times 10^{24}$ kg.

The mass of Jupiter is $(9.55 \pm 0.08) \times 10^{-4}$ times the mass of the Sun.

The mass of Saturn is $(2.72 \pm 0.06) \times 10^{-4}$ times the mass of the Sun.

The mean density of the Sun is 1.42 ± 0.01 g/cm^3.

The Surface Temperature of the Sun is a bit more than 5100 K.

Notice that I have tried to include some sort of estimate of the uncertainty in each value. The line "The Sidereal Month is 27.3214 ± 0.0004 days" means that the value probably lies between $27.3214 - 0.0004 = 27.3210$ and $27.3214 + 0.0004 = 27.3218$ days. For a statistical result such as this one, there are standard formulas for calculating the uncertainty. In other cases the value depends much more on the calibration of the equipment used, and, when I get desperate, it depends on a gut feeling! It is normal scientific practice to include an estimate of the uncertainty with any measurement. I have included an appendix with some ideas about error analysis. However, remember that we are doing this because it is fun. If this bit of the analysis leaves you cold, then don't do it. Be that as it may, the result of a measurement has much more meaning if it does include an idea of the uncertainty. For example, my value for the radius of the Earth actually worked out to 4208 miles. But I estimated the uncertainty as 6%, or about ±250 miles, so it is actually misleading to give more figures than I have included in the list above. Perhaps a more telling example is that, in chapter 17, I calculate the ratios of period squared divided by distance cubed for each of the planets. If my measured values were exact, Kepler's third law predicts that the ratios should come out equal. Because the values contain experimental error, the ratios do not come out exactly the same. But because I have kept track of the uncertainties I can claim that my numbers are consistent with the ratios being equal. I do the same thing for Jupiter's moons and then use the value of the ratio to find the mass of Jupiter. For the surface temperature of the Sun, by the nature of the method used, my number is a lower estimate. It is actually about 10% below the handbook value. As I describe in later chapters, the uncertainties for the sidereal periods of the planets are statistical uncertainties only and there may be additional errors because I assumed in the analysis that the planets move in circular orbits. My measurements of the tropical year and the inclination of the ecliptic to the equator are described in Chapter 8 at the beginning of Part II. The measurement of the eccentricity of the Earth's orbit is described in chapter 21. Most of the time measurements are described in Part I, the distance measurements are in Part II, and a few measurements of physical properties of the Sun and planets are described in Part III.

I can use some of these numbers to verify some bits of the clockwork. In particular, see Chapter 17, "Did Copernicus and Kepler get it right?" Alternatively, I could get more accurate values in some cases by assuming the correctness of, for example, Kepler's third law. The numbers given in the list above, such as the synodic and sidereal periods of the planets and their distances from the Sun, are all independent measurements. If I assume Kepler's third law, I can take the most precise of the measurements and use it to find better values of the others. I explore this line of attack in Chapter 18.

I am missing one big one in all of this. I haven't yet attempted to measure the astronomical unit, the distance of the Earth from the Sun, so that all the distances I measure outside of the Earth - Moon system are relative ones. An article in Sky and Telescope, January 2007, pp 91-94, by Robert Vanderbei and Rus Belikov, shows that the measurement can be made with equipment only slightly more elaborate than I have used, so maybe I shall attempt it yet! If I look up the value of the astronomical unit in a handbook, I can rewrite some of the entries in the list as:

The diameter of the Sun is $(1.389 \pm 0.007) \times 10^6$ km.
The diameter of Mercury is (5000 ± 600) km.

Gauging the Solar System

The diameter of Venus is (12400 ± 300) km.
The diameter of Mars is (7000 ± 1000) km.
The diameter of Jupiter is $(1.44 \pm 0.03) \times 10^5$ km.
The diameter of Saturn is $(1.2 \pm 0.1) \times 10^5$ km.
The mass of the Sun is $(1.9885 \pm 0.0001) \times 10^{30}$ kg.
The mass of Jupiter is $(1.90 \pm 0.02) \times 10^{27}$ kg.
The mass of Saturn is $(5.4 \pm 0.2) \times 10^{26}$ kg.

A second number that I use but haven't measured for myself is the universal gravitation constant, G. I did actually borrow a student version of the apparatus to measure this constant from my old department, but I concluded after working with it for some time that I was not going to get a value of anything like the precision I needed so I might as well bite the bullet and use the handbook value. Of all the fundamental constants, this is the one that is known to the smallest number of decimal places and also the experiment is very sensitive to outside vibration.

I don't expect ever to run out of projects of this sort. New ones keep appearing. In Chapter 16 I describe some observations that showed me a way to measure the draconic month, and in Chapter 21 there are measurements that lead to a value of the anomalistic month. Neither of these has the precision of my measurements of the synodic and sidereal months, as yet, although they are good enough to show the difference between these slightly obscure periods and the sidereal month. I did include them in the table of results but I regard them as works in progress. Perhaps, also, I need to improve my technique for photographing through a telescope. My results on the angular sizes of the planets are very incomplete and there are fascinating conclusions to be drawn from those measurements.

The second edition of the book mostly records a gradual improvement in the precision of my values. There are more substantial additions in Chapter 13, describing the retrograde motion of Venus, and in Chapter 26 on variable stars. Chapter 22 now includes a measurement of the mass of Saturn. There two new chapters, 15 and 18, which have expanded from single paragraphs in the first edition. Chapter 15 includes observations of the asteroid Vesta, and Chapter 18 contains a set of planetary parameters based on the assumption that their orbits are exactly elliptical.

As a final comment, I have written this book more or less in the form of a diary, documenting how I figured out how to measure the various numbers. What I am trying to say is not "Look what I've done", but "Look what you can do!" Have fun.

Part I. Solar system time measurements

Chapter 1: One to get started - New Moons

New moons and Old moons

Looking for new moons is an easy way to slide into learning about the "clockwork" of the solar system. This is a good subject to talk about if you are making a presentation to a non-astronomical audience. Most people have never seen a moon newer than a couple of days, and many of them do not realize that there is one moment in time that is the point of new moon. Also many are not aware that there is a thin crescent old moon before sunrise just prior to a new moon. A thought provoker is to show photographs of a new moon and of an old moon and ask what is the difference? Looking for very new moons may have little scientific value but it is a good game. I always get a kick out of seeing a new moon less than thirty hours old, or an old moon less than thirty hours from new.

Two sources of frustration are that you have to accept the times that nature gives us - you can't simply say "today I shall look at a 24-hour-old moon", and, when the clockwork does oblige and there is a 24-hour-old moon to be looked at, the weather may choose not to cooperate and you have to wait for next time. Anyway, let's start by looking at an almanac for 2010. All the times here are standard time for the East Coast, EST, and for a latitude of 40° North.

Date and time of new moon	Time of sunrise	Time of sunset	"Age" of old moon on previous day (hrs)	Age of new moon on next day (hrs)
Jan 15, 2:11	7:20	16:59	19.4	39.3
Feb 13, 21:51	6:55	17:33	39.4	20.2
Mar 15, 16:01	7:12	18:07	34.3	26.6
Apr 14, 7:29	5:23	18:37	26.6	35.6
May 13, 20:04	4:47	19:06	39.8	23.5
Jun 12, 6:15	4:31	19:29	26.2	37.7
Jul 11, 14:40	4:41	19:30	34.5	29.3
Aug 9, 22:08	5:06	19:04	41.5	21.4
Sep 8, 5:30	5:34	18:21	24.4	37.4
Oct 7, 13:44	6:02	17:33	32.2	28.3
Nov 5, 23:52	6:35	16:53	41.8	17.5
Dec 5, 12:36	7:06	16:35	30.0	28.5

I make a chart like this for myself at the beginning of every year.

The first three columns are taken from the Observer's Handbook. The last two columns are calculated allowing thirty minutes for the sky to darken. By "age" of the old moon, I actually mean the time before the moment of new moon. I started looking for new moons with just about this much guidance and I initially had quite a few fruitless evenings. One problem is that the age of the moon is not a complete indicator of how high in the sky it will be. The Moon moves away from the Sun at an average rate of about 12° per day. (It moves round its orbit at an average rate of 13° a day, but the Earth moves round the Sun at 1° a day.) A one-day-old moon would be easy to see if it stood 12° above the horizon at sunset, but its separation from the Sun is measured along the path of its orbit. This coincides approximately with the ecliptic, and the angle the ecliptic makes with the horizon varies through the course of the year and even through the course of each day. The graph below shows the angle between the horizon and the ecliptic at sunrise (dashed line) and at sunset (solid line) for the latitude of 40° North (there is a very slight variation from one year to the next). I calculated these using formulas in Jean Meeus's book "Astronomical Algorithms".

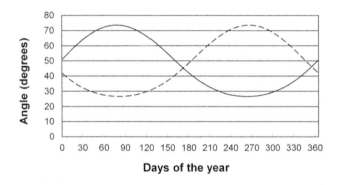

Figure 1.1 Angle between the horizon and the ecliptic

In the worst cases, where the angle is less than 30°, a one-day-old moon would be less than 6° above the horizon at sunset, and probably have set by the time the sky was dark enough for you to see it. Broadly speaking, the first six months of the year are the best time to look for new moons in the evening, with the middle of February to the middle of April being the very best time, and the last six months are the best time to look for old moons in the morning, with September and October being the best period. Based on these ideas, the most promising entries in the table are probably the new moons in February and March, and the old moon in September. I did, in fact, get to see the new moon in March and the old moon in September. The February new moon was rained out.

There are more factors involved. You may have noticed that I used the words "approximately" and "average" in describing the path of the Moon. The Moon doesn't move exactly along the ecliptic. It might be slightly to the North of it, or to the South. If it is to the North, then viewing in the Northern hemisphere is improved, and if it is to the South, the chances are improved for someone in the Southern hemisphere. In fact, it can be as much as 5° above or below the ecliptic. I came to appreciate this variation when I failed to see an old moon I was hunting in three consecutive months in the late summer of 2014 and realized that at least part of the difficulty was that the Moon was below the ecliptic for each of them. This pattern slides backwards on the calendar from one year to the next, repeating with a period of 18.4 years, but at some point in any year there will be a group of a few months that are more favorable than average and a group that are unfavorable. I have come to grips with this now. In Chapter 16 I describe measurements of the angular distance the Moon is above or below the ecliptic and I can use those to predict quite accurately what to expect for a particular new or old moon.

Also, the Moon doesn't move around its orbit at exactly the same speed each day. It travels in an approximately elliptical orbit and its speed round its path varies, also approximately, in accordance with Kepler's second law. When the Moon is closest to the Earth, at perigee, it moves more than 14° each day whereas when it is furthest from the Earth, at apogee, it moves less than 12° each day. This was easy to see in the new moon of March 15, 2010. On the evening of the 16th the Moon was about 26 hours old, but I and several of my club members noticed that it was clearly narrower than we expected. My friend Joe Stieber pointed out that the Moon was close to apogee. There is nothing you can do about these small effects, but it is good to be aware of them and to notice any variation from expectations.

There is a number of 7°, known as the Danjon limit, for the minimum separation of the Sun and the Moon such that the Moon is still visible. Danjon based this on the idea that the amount of light reflected into the crescent by the slightly rough surface of the Moon drops sharply at this separation. There is some dispute about whether the effect is instead a psychological one. There is a good discussion, with references, on a web page (in 2016) http://the-moon.wikispaces.com/Danjon+Limit. A personal problem I have with this is that I need to consult an almanac or a planetarium program to find the separation of the Sun and Moon. This is a bit of a stretch from a do-it-yourself viewpoint.

A way to add interest to new or old moon photographs is to wait until a planet is close to the Moon. Figure 1.2 shows a photograph of the new moon and the planets Venus and Mars that I took on February 20, 2015. Notice the Earthshine that illuminates the dark part of the Moon.

Figure 1.2 New moon with earthshine, Venus and Mars. Mars is the upper planet.

The bottom line is that the newest moon I have managed to see so far was 20 hours and 47 minutes old, on April 25, 2009. I saw that through binoculars only. The newest moon that I have been able to see naked eye was 22 hours and 40 minutes old, on February 28, 2006. My eyesight is not great, so you can do better than this!

Old moons and New moons

Here is another challenge. Can you see the old moon and the new moon of one cycle? If so, how close in time would they be? A quick answer is that you can see an old moon about 24 hours before new, and a new moon about 24 hours after new, so 48 hours, or two days, seems to be the answer. A second thought shows that this is wrong. You view the old moon in the morning and the new moon in the evening, so the interval between them must involve a half-day. It must be one-and-a-half days or two-and-a-half days or even three-and-a-half days. The graph of the angle between the ecliptic and the horizon plays an important part in planning this, since you need to be able to see both old and new moons at almost the same time on the graph. The period around the middle of the year works best, although the end of the year period can lead to a shorter time between the observations. The weather also has two chances to spoil things. It is somewhat irritating if you get to see the old moon one morning and then the weather turns bad for the next few evenings. I was able to see old and new moons separated by three and a half days several times before I managed a two-and-a-half day interval. Figure 1.3 shows two photographs taken 5:11 am EDT on May 23rd, 2009, when the moon was just 27 hours from new, and 8:48 pm on May 25th, 2009, when the moon was 36 hours and 37 minutes old, an interval of 63 hours and 37 minutes.

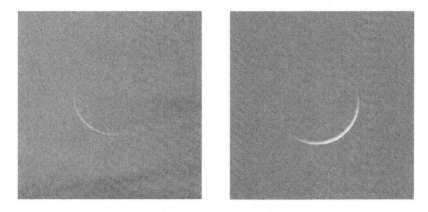

Figure 1.3 Old moon and new moon

I have repeated the double observation several times and my current best time interval between the two sightings is 59 hours and 48 minutes. To put this in perspective, the October 2008 issue of the Sky and Telescope Magazine carried a letter from Roy Hoffman, of the Israeli New Moon Society, reporting that on June 3rd and June 4th of that year members of the society had seen and photographed the old moon and the new moon on successive days. Wow! The interval between the two sets of observations was slightly less than 40 hours. They did make their observations at the best time of year with respect to the angle of the ecliptic, and in a year when the new moon was almost at its maximum angle above the ecliptic in June. So I have something to aim for, but I am not holding my breath.

It should be obvious that you need observing sites with both good eastern and good western horizons in order to observe both old and new moons. I have neither from my backyard so I need to make a short excursion in either case. If you are visiting a new site, it is worth taking a magnetic compass along to establish just where the east and west directions are. There is a bonus to using the same location to observe the old or new moons from for the whole year. It may come as a surprise to see how far north or south of the nominal direction the moon actually rises or sets. The same comment applies to the Sun. If you view sunrises from the same location at the Winter and Summer Solstices and at the Equinoxes, the variation in direction will probably come as a surprise. Figure 1.4 shows a composite of photographs taken just after sunrise on the Summer solstice, the Fall equinox and the Winter solstice, all from the same point. The zoom lens was set to a focal length of 18 mm, giving a 67° field of view. The photograph taken at the Spring equinox had the sun virtually on top of the Sun at the Fall equinox.

Figure 1.4 Sunrises through the year

Chapter 2: Synodic Month and Sidereal Month

I have two bits of advice to anyone who decides to carry out a similar program of measurements to mine. The first is to keep a log book and to make it much more detailed than you think you are going to need. In the pages that follow, I several times regret that I didn't record some bit of information that I would now like to have. The second is to work from the beginning in Universal time. In several cases, I had to go back over old records and convert them, because I found belatedly that it was easier to work with universal times and dates.

The synodic month is just the (average) time from new moon to new moon. This is one of the numbers that got me started on this whole set of projects. One evening in March, 1987, I stepped out of my front door, looked up in the western sky, and thought "Wow, that's a really new moon!" In fact it wasn't that new. I later recognized that it was between two and three days old. At the time, I was sufficiently impressed to make an entry in my star log, and I later went back and put a number "0" by it. Since then, I have been noting down the first time each month that I see a new moon, and the numbers have passed 400. (There are several less personal numbering systems. The best known is the Brown lunation number, which starts with the first new moon of 1923.) Actually, I now try to get the date when the moon is again between two and three days old. With practice you can learn to recognize how wide the crescent should be. Getting a good value for the average length of a month takes a little thought. The simplest approach is just to divide the number of days from the first observation by the counter number (remember that the first new moon was numbered zero). This doesn't converge particularly well. In 2019, thirty-two years after I started accumulating the data, the running average varied between 29.528 and 29.533, whereas I am trying to narrow down the fourth decimal place. There is a reason for this slow convergence. Most measured quantities that include random deviations from a mean show a Gaussian distribution around the center - the famous bell-shaped curve. I have used a computer to generate the distribution of new moon lengths over a long period of time. I used one of Duffett-Smith's subroutines to find the lengths of 20,000 months, sorted them into bins with a width of 0.01 days and plotted them as a histogram. The result, shown in Figure 2.1, doesn't look much like a bell. (I later found a more accurate calculation at eclipse.gsfc.nasa.gov/SEhelp/moonorbit.html, fig. 4.3) You can see that the interval from one new moon to the next can vary by as much as 0.3 days from the average value, and that the distribution has two maxima. This double peaked type of distribution arises when there are oscillatory corrections to the mean. It is the presence of these that slows down the convergence of the running average. Over a long enough time interval the oscillatory terms must average to zero, but in the case of the

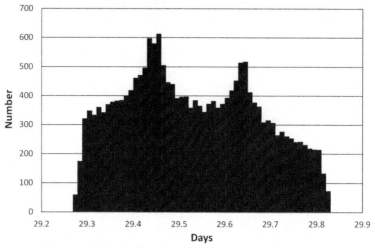

Figure 2.1 Distribution of lengths of synodic month

month it takes a really long time. A better way of pulling out the mean value from the observations is to plot the dates of the new moons against the number in a spreadsheet and to fit a straight line to the points on the graph. The straight line smooths out the oscillations to some extent. What I actually plotted along the vertical axis was the number of days since that first new moon #0. I plotted the points in Excel and asked for a linear trend-line with the equation displayed on the graph. This gave me the result shown in Figure 2.2.

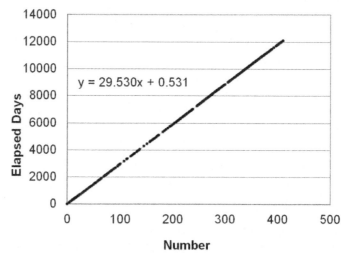

Figure 2.2 Observations of New Moons

The fitted value of the synodic month is the first value in the equation, 29.530 days. (This is the first of many times that I use this sort of analysis. Try it for yourself if you aren't familiar with it.) If you are thinking that you don't intend to start on a project that will take 34 years to complete (if, indeed, it is now complete), then take heart. The first three digits, or 29.5, emerge clearly after one year of observation. The next digit does take eight to ten years to establish itself. If that factor of ten sets a pattern, I am unlikely to get the next digit completely pinned down! Excel also knows the formula for the uncertainty in the slope, but it needs a slightly complicated array function LINEST( ). At present, it gives an uncertainty of 0.0006 days, which is just under a minute.

There is another way to find the length of a synodic month that might occur to you, especially if you have been observing the sky for some time, and that is from the dates of eclipses of the Sun. An eclipse surely pins down the moment of new moon. I mention in chapter 23 that I took a pinhole photograph of a partial eclipse on March 7, 1970. That was around noon. Much more recently, I observed a partial eclipse at sunrise on November 3, 2013. The time between these two events was 15946.8 days. I know that the length of a month is about 29.53 days. 15946.8 divided by 29.53 is 540.02. This must be a whole number of months, so I can round it to 540. Now I divide 540 into 15946.8 to get a better estimate of the synodic month of 29.5311. This trick, of dividing a long time interval by an approximate period, rounding to a whole number, and then dividing the long interval by that whole number is very useful; I use it several times in later chapters. The uncertainty in the time interval is about 3 hours, and this divided by 540 gives an uncertainty of 0.0002 days. This is a very precise value for the average synodic month over this particular 43½ year interval, but the difference between this average and the handbook value is several times the uncertainty. 43½ years is still not enough time for all the oscillatory terms to average to zero. There is also the objection that I rely completely on other sources to tell me when an eclipse is going to occur, so perhaps we count this as cheating!

A second type of month is the sidereal month. This is the time it takes the Moon to complete an orbit round the Earth, measured relative to the stars. This is slightly shorter than the synodic month, which is the time from one new moon to the next, because of the motion of the Earth round the Sun. My method for finding a sidereal month is quite similar to what I did for the synodic month. I keep a month-by-month record of the dates when the Moon reaches a particular position with respect to the stars. Of course, I need a well-marked position in the sky and I chose the star Spica, which is

17

reasonably close to the ecliptic. Looking back, I would probably have done better to use Regulus, which gets higher in the sky in my part of the world, but that didn't occur to me until I had at least ten years of observations invested in Spica. If you try this, you might choose more than one star and get several parallel sets of data. The procedure is to watch as the Moon approaches Spica, night by night. When it gets within about 20° or 30°; make an estimate of the angle between the Moon and Spica. Keep that up until the Moon has gone past the star and you can then interpolate to find a good estimate of the time when they were closest. I find that I can make the estimate accurate to a few hours. Since this is much smaller than the variations in the synodic month, I initially thought that this would make it easier to extract the average time, but, like the synodic month, the sidereal month is not exactly the same length each time and there are oscillatory corrections to be averaged out. So I finished up plotting a graph in Excel and extracting a trend line, just as I did for the synodic month. Figure 2.3 shows the result. There are fewer points in this plot because I only get to see Spica for part of each year and I need several clear nights in a row to make a good interpolation.

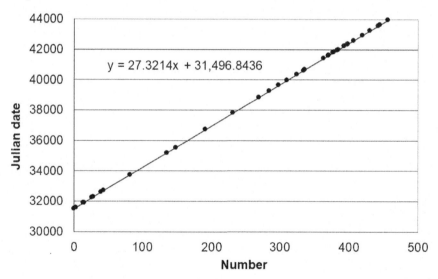

$$y = 27.3214x + 31,496.8436$$

Figure 2.3 Conjunctions of Spica with the Moon

What is plotted vertically this time is the Julian day number. This is a numbering scheme for the days that bypasses the complications of leap years and differing month lengths. I used a program from the book "Astronomy with your personal computer", by Peter Duffett-Smith, to calculate these. They are values of what he calls DJD, or Dublin Julian Day, which starts the count at the beginning of the twentieth century. The formulas are also in "Astronomical Algorithms" by Jean Meeus. The fitted slope is 27.3214 days, with an uncertainty of 0.0004 days, or 36 seconds. This set of numbers does seem to be converging a little better than the values of the synodic month, and in Chapter 17 I use this value to get a more precise number for the synodic month. As was the case for the synodic month, it will take you over thirty years to reproduce this curve, but surely the most interesting result is to verify that the sidereal month is more than two days shorter than the synodic month, because of the motion of the Earth – Moon system round the Sun, and this emerges after only a few months of observations of the two sets of data.

Chapter 3: Sidereal Day and Sidereal Year

A good question for presentations to the general public is to ask "What is a day?" The commonest answer is often just "Twenty-four hours", which doesn't really get to the point. With a little prompting you will probably get two more thoughtful answers. The first is usually the time it takes the Earth to turn on its axis, and the other will be along the lines of the time from one noon to the next. The two are slightly different, because the Earth moves round the Sun at the same time as it turns on its axis. Historically, we set our clocks by the Sun, so the noon-to-noon time is what we call 24 hours, or one day, and the time for the Earth to turn on its axis is about 4 minutes less. From an astronomical perspective the impact of this is that the stars rise 4 minutes earlier each day, which adds up to 2 hours each month and a full 24 hours over a year, getting the night sky back to its starting point.

I adopted a very simple technique for measuring this 4 minute advance of the stars more accurately. Since I live at a latitude very close to 40° North, and the bright star Vega has a declination of 38° 40', it follows that Vega passes over my house each night during the summer months. There is a brick chimney at the east end of the house. I found that if I scrunched my face up against the side of the chimney I could see a rather sharp moment when Vega was occulted. I carried a stopwatch which I started at that instant. Then I walked inside the house and compared the watch with an accurate time signal from short wave radio (this was in the 1980's; in the 21st century I could probably do all of the timing using a smartphone). I kept this up for five summers, from 1987 to 1991.

My method of analysis will sound familiar if you have read the previous chapter. I plotted the times of occultation versus a numerical counter of the passage in Excel and asked for a linear trend line. It does take a little care to get the points for different years in the right place. Hopefully, in the first year you made enough observations to give you a good first approximation to the sidereal day. Then, when you measure the first occultation of the second year, you can divide the time interval since the last observation of the first year by this value to tell you how many times Vega has gone overhead in between. And the same thing applies to subsequent years, except that your estimate of the sidereal day is steadily improving. I did make more observations in the first two years than later on. My results are shown in Figure 3.1.

The slope of the final plot gives the length of a sidereal day as 0.9972697 solar days. This translates as 23h 56m 4.10s, which is 3 minutes and 55.90 seconds short of 24 hours. The uncertainty is ±0.01 seconds. The number I have measured actually corresponds to what the Observer's Handbook calls a Mean Rotation Day. The handbook defines a sidereal day in terms of the equinox to equinox time. The difference between the two arises from the effect of precession of the Earth's rotation axis, and is smaller than the uncertainty of my measurement. I had hoped that my measurements would each be accurate to one or two seconds. The average discrepancy between my measured points and the fitted straight line is 10 seconds, and I take this as a measure of the actual accuracy of my numbers. Perhaps the extent to which I scrunched up my face varied from night to night!

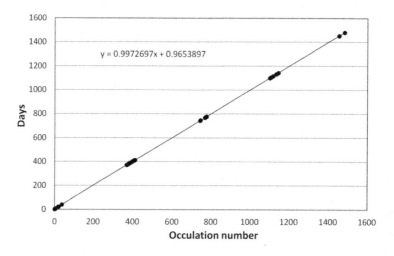

Figure 3.1 Occultations of Vega by the roof of the house

I have a confession to make. When I gave up making these measurements, I had it in mind to start again in maybe twenty years' time, to give myself a very long baseline for the graph. Well, I tried it. I observed several occultations of Vega in the summer of 2009 and added them to the graph. The problem was that they clearly didn't fit on the same line! The discrepancy was minutes rather than seconds. I checked everything I could think of, and of course a lot had changed in twenty years, including a new roof on the house. Finally, an architect friend of mine was visiting and I poured out my troubles to him. He eyed the chimney and argued that it could easily have shifted by several degrees when the roof was changed. A tilt of 1° would change the occultation time by four minutes. So for now I am going with the old measurements. I regretted that, because I had a good reason to try to reduce the uncertainty in my measurement, described in the next paragraph. There is a moral here. If you are planning to start a similar series of measurements, make sure you choose a reference marker that will stay in the same place for a long enough time!

I can use the last value, that a sidereal day is 0.9972697 solar days to find the length of a sidereal year. The formula is that, if an orbit of the Earth round the Sun lasts for N solar days, then it lasts for $(N+1)$ sidereal days. Picture a long piece of string connecting the Earth to the Sun. Every time the Earth turns on its axis the string wraps once round the equator. But as the Earth moves round the Sun it tends to unwrap the string so that in a complete revolution round the Sun one of the turns of string is completely unwound. Therefore

$$(N+1)\times 0.9972697 = N$$

$$\therefore N = \frac{0.9972697}{1-0.9972697} = 365.26 \pm 0.01.$$

The solution of this is N=365.26 (solar) days. The uncertainty is actually a bit less than 1 in the second decimal place. In chapter 8 I describe the type of measurement that I used to find a value for the tropical year, and at present the uncertainty in my best value is again about 0.01 days. I think the difference between the sidereal and tropical years is my best chance for getting any sort of estimate of the rate of precession of the Earth's rotation axis. The actual difference between the two years is approximately 0.014 days, which corresponds to a period of 26,000 years for the Earth's axis to make a complete cycle. As long as the uncertainties in my measurements are 0.01 days, I can say nothing about this.

Chapter 4: Synodic periods of the inferior planets

The synodic periods of the planets are their repeat periods as seen from the Earth. The outer planets, Jupiter, Saturn, Uranus, and Neptune, move much more slowly than the Earth, so their repeat appearances are mostly caused by the Earth's coming round to them and their synodic periods are only slightly over a year. For Mars, and the two inferior planets, Mercury and Venus, the periods are more a result of the combined motions of the two objects. I have been keeping a log book of my astronomical observations, however casual, for over thirty years, and my first thought was to extract as much information as I could from this record. It worked out pretty well, but I realized that my records weren't detailed enough. If you don't keep a log book, I strongly recommend that you start one and that you make it better than mine! In particular, if you are looking at a planet, record at least the constellation it is in or the stars it is near, and the time of evening or night.

My analysis was different for the inferior planets, Mercury and Venus, than for the superior planets, Mars, Jupiter, Saturn and Uranus. I have records of a few telescopic sightings of Neptune, but so few that I have not attempted to include them. I do regret now that I gave up on Neptune so quickly!

In the cases of Venus and Mercury, my first step was to list in a long column all of my sightings of each, together with a notation of whether they were evening or morning sightings. Actually, all I needed were the first and last days I had seen the planet during a particular evening or morning appearance. I also calculated the Julian day for each observation. For Venus I had another important datum. I did take a trip to England in 2004 so that I could watch the transit of Venus in front of the Sun in its entirety, so I had a very accurate value for one inferior conjunction. I was not able to see the transit of 2012. The pattern is that an inferior planet shines in the west in the evening sky, moving out to its greatest eastern elongation from the Sun. It then moves between the Earth and the Sun passing through inferior conjunction, when it is as near as possible to being in line between the Earth and the Sun. Then it appears in the east in the morning sky, heading out to greatest western elongation, and then swings behind the Sun passing through superior conjunction. The part of the orbit from greatest eastern elongation to greatest western elongation is shorter in distance round the orbit, and takes less time, than the part from greatest western elongation to greatest eastern elongation. It is also true that, while I have talked of seeing Venus in the evening or morning, it is sufficiently bright that it is sometimes possible to see it while the Sun is in the sky.

Venus was slightly easier to analyze than Mercury because I had the date of an inferior conjunction. The method is to make a guess at the period and to count backwards and forwards in the table from the known date of inferior conjunction, in steps of half a period, marking in where the change from one type of elongation to the other occurs. If the guess is a bad one, the changeover will eventually disagree with what was observed. By trial and error, guessing different values for a period, I found that my observations were consistent with only a narrow range of values. I could pin the period down to 586 ± 4 days, better than 1% precision. A single evening observation in 1988 turned out to be a key to fixing the upper limit and an observation of a very thin crescent Venus in 2009 fixed the lower limit. The handbook value is 583.9 days.

The analysis for Mercury was similar except that I didn't initially have a firm date for inferior conjunction. I found a place in my column of sightings where there was the shortest gap between evening and morning observations. When I started this analysis, it was a 33 day gap in Jan-Feb 2008, though I have narrowed it down since then. By a double trial and error, I first guessed a date for the conjunction and then tried different values of the period, and repeated the process for several dates for the conjunction. It turned out that my data pinned the period down very well, even allowing a range of dates for the conjunction. It helped that the period for Mercury is much less than that for Venus so I had more cycles to work with, 98 as opposed to 21 for Venus, in 2020. My observations are consistent with periods only in the range 115.9 ± 0.3 days. I started doing these calculations on paper, but quickly switched to a small Basic computer program. I have also set it up in Excel. If you are happier working with spreadsheets you can do it that way.

I was pleasantly surprised by these results and assumed that the answer would improve as I continued making observations, and particularly if I made an effort to see the planets closer to the inferior conjunction. For Venus, there was a favorable pair of eastern and western elongations in December 2013 and January 2014, favorable because Venus was well to the North of the ecliptic so that it was visible from the Northern hemisphere relatively close to conjunction. I was able to see Venus on the evening of January 7, 2014, and then on the morning of January 17, 2014. The two observations were less than ten days apart. For comparison, at the next inferior conjunction, in 2015, my closest naked eye observations on either side of the conjunction were just over 31 days apart. I had seen that coming. The synodic period of Venus is roughly equal to two and a half times its sidereal period. If Venus is well to the North of the ecliptic at one conjunction, it will be well below it at the next. This isn't exactly the case, of course, or the rare event of a transit of Venus would never occur. I watched another close pair of observations in 2017. If I include the four dates from 2014 and 2017 in the spreadsheet they limit the acceptable values of the synodic period to 584.1 ± 0.6 days.

In the case of Mercury, however, I eventually realized that I had been rather lucky. The problem is that the time from one inferior conjunction of Mercury to the next one can vary quite wildly, presumably because the orbit of Mercury is by far the most elliptical of all the planets. The time can be as much as twelve days longer or shorter than the average interval. My program takes uniform time steps of one half of the guessed synodic period to locate the change-over points between eastern and western elongations. If I had been able to observe Mercury within just a few days of inferior conjunction, my computer program would have found that there was no value for the synodic period that reproduced all the data. The naïve method of analysis I used is ideal for data that are concentrated around the maximum elongations but spread over as many years as possible. Of course, if you do have observations close to the conjunctions, you can always leave them out for this stage of the analysis and then use them to tell you something about the ellipticity of the orbit. Venus has a much less elliptical orbit so that this did not cause a problem. Actually, I can find from handbooks that individual times from one inferior conjunction to the next can vary by three or four days from the average for Venus. This means that I am just about at the limit of what I can get from this technique. I suspect that a primary reason for this in the case of Venus is that the Earth itself does not move in a smooth ellipse. Instead the center of mass of the Earth-Moon system moves smoothly, and the Earth wobbles around this point. Another result of this is that Earth reaches its perihelion point at slightly different times in successive years. The precision of my measurement is getting higher than the accuracy of my theory. (I never really came to grips with the difference between precision and accuracy before this!)

While these results are not bad, I thought I might be able to do better, at least in the case of Venus, using the results from the photographic methods described in Chapter 13 to pin down the inferior conjunctions more precisely. In order to find the size of the orbit of Venus, I measured its position relative to the stars at an eastern elongation in the summer of 2010 and at the following western elongation early in 2011. As a side result, I found the dates of the two maximum elongations. The half-way point between these was 10/29/2010, 0h 14m Universal time. This should be very close to the time of inferior conjunction; there could be a slight difference because of the ellipticity of the orbits. I have since repeated this set of measurements twice more. Also as I mentioned, I watched the whole of the transit of Venus in

Time measurements

2004. I estimated the center of the transit to occur at 8:20 am UT on 6/8/2004. I therefore have observed or estimated times for four inferior conjunctions and I can use my usual analysis procedure of plotting a line in Excel and asking for the fitted slope. It comes out as 584.0 ± 0.25 days, so this is my preferred value for Venus' synodic period.

Both the methods I have used for Venus require observations spread over several years. In his book "Do-it-yourself Astronomy", Sidney G. Brewer gives a third method that needs only about six months of observations. It builds on the same measurements of eastern and western elongations that I used in my second method. You also need to measure the maximum angle that Venus moves away from the Sun in each case, but that was the main point of the measurements in the first place. Briefly, it works like this. As I report in Chapter 13, I found that Venus reached a maximum eastern elongation on 8/19/2010, at 7:12 Universal time, DJD 40407.8. It was then 46.0° east of the Sun as seen from the Earth. Equivalently, the line joining Venus to the Sun made an angle of $(90.0 - 46.0) = 44.0°$ from the line joining the Earth to the Sun (take a look at figure 13.1 if this is not clear). I also found the following maximum western elongation was reached on 1/17/2011, at 17:17 UT, DJD 40549.2, and it was then 47.0° west of the Sun. The line joining Venus to the Sun therefore made an angle $(90.0 - 47.0) = 43.0°$ from the line joining the Earth to the Sun. (Both the angles are good to a tenth of a degree. It just happens that the last significant figure is zero in both cases.) The time interval between the two maximum elongations is 141.4 days and, relative to the moving Earth, Venus has moved $(44.0 + 43.0) = 87.0°$ round its orbit. The synodic period is the time for it to move 360°, relative to the Earth, and this is

$$141.4 \text{ days} \times (360° / 87.0°) = 585 \text{ days}.$$

This doesn't seem to be quite as accurate as my two methods, but it needed fewer observations. It does involve the assumption that Venus moves at a constant angular speed, and this is probably the factor limiting the accuracy.

I haven't managed to apply these photographic methods to Mercury yet. The difficulty is that I need to be able to follow Mercury over a period of nights, and take photographs that show identifiable stars. So far, I have cheated to the extent that I look up the dates of maximum elongation in the handbook. So the value of 115.9 ± 0.3 days that came from my old log books is still the best I have. Mercury made a transit in front of the Sun on May 9, 2016. Knowing that it was to happen I tried to predict for myself the date it would occur. The best that I could say from my numbers was that it would be in the range of May 8 to May 18. This did bring home to me the difficulties of astronomers of three hundred years ago in predicting transits of Mercury and Venus.

I did in fact manage to see the transit of Mercury, which gives me a very accurate time for an inferior conjunction. Actually, I saw two of them, one on May 9th, 2016 and a second on November 11th, 2019. The 2016 transit is shown below. Mercury is the lower black dot.

Figure 4.1 Transit of Mercury on May 9th, 2016

When I used either of these dates as a starting point for my analysis the result was very little changed from the initial value of 115.9 ± 0.3 days. Also, the two transits were too close together to give an accurate time for the synodic period based on the time between them.

Chapter 5: Synodic periods of the superior planets

For the superior planets my analysis was different. In Chapter 14 I describe how I can pin down the times of opposition of the superior planets to an accuracy of about one hour. If the planets moved in circular orbits at constant speeds, their synodic periods would just be the time intervals between successive oppositions. However in reality the interval between any two successive oppositions can differ from the long-term average by several percent, and I have no simple way to make corrections. So once again, I started by getting what I could from my old log books. The difficulty here is that we are viewing the other planets from a moving platform, the planet Earth. So I had to concentrate on observations taken when the planets were in identifiable positions in the sky. I found that many of my records were less helpful than they might have been because I hadn't recorded enough information about the part of the sky where the planet had been. I was most consistent in noting down when I had seen the planets low in the western sky just after sunset, so I made use of these. For each planet I plotted a graph of the date of such a sighting versus a counter of times the planet had come around, and, as usual, asked Excel for trend lines. Jupiter and Saturn come around pretty well once a year, so the counter goes up by one each year; Mars comes round roughly every two years. The ancient Egyptians looked for the first sighting of the star Sirius in the morning sky just before sunrise, called the heliacal rising. I think of these observations as the heliacal setting of the planets. The values that currently come out of the plots are 786 ± 6 days for Mars, 399 ± 1 days for Jupiter, and 378.9 ± 0.4 days for Saturn. A problem with this method is that the earlier observations were made quite casually whereas in more recent years I have been making a special effort to see the planets low in the west. In consequence, the Excel plots have been curving upwards. If I can keep up a consistent set of observations for several more years the curves should straighten out again. The uncertainties are based simply on the scatter of points in the graphs and do not allow for my observational inconsistency! The heliacal setting is as close as I can get to seeing the planet at superior conjunction but I should point out that a factor that may limit the accuracy of this method is that the synodic periods of the planets are not simply related to the Earth's year, so that the heliacal setting time comes at a different date each year, and therefore at a different time of the evening as the time of sunset varies, as well as a different point on the horizon.

A potentially better way to pin down the synodic periods of these planets is to locate the dates of oppositions spaced as many years apart as possible. I have been improving my technique at this in order to use the information to calculate distances to the superior planets. That is described in Chapter 14. I have pretty good data for the last twelve years and some very spotty observations before that. As I just pointed out, it is not good enough simply to observe two successive oppositions and take the time difference. The difference does vary by a few percent because of the elliptical shapes of the orbits of the Earth and the other planets, and also because the Earth moves in a slightly wobbly path around the center of mass of the Earth – Moon system. For example, I located two successive oppositions of Mars on December 25, 2007, and January 30, 2010. The interval between them is 767 days. If I compare with the handbook, both of my dates and the interval are quite accurate, but the handbook value for the synodic period is 780 days. The discrepancy is about 1.7%. I believe the slightly lower value in this case came about because Mars was near its aphelion position and traveling relatively slowly, but I simply do not have enough data on Mars to confirm this. This illustrates a rather general pattern. The more precisely you try to measure a particular value, the more you need to worry about small effects.

For Mars I had made one earlier estimate of an opposition. I watched it in retrograde motion in November and December of 1990, and estimated the date of opposition as December 5, 1990. I can combine this with the accurate dates for the five more recent oppositions listed in Chapter 14. A graph plotted in Excel has a slope of 776 days. Even if my early observation was in error by a whole month, this value would change by only 3 and a bit days, so I prefer this number to the heliacal setting value and go with 776 ± 3 days. In situations like this, I wish I could go back in time and redo the old observation!

For Jupiter I have pinned down oppositions on August 14, 2009, September 21, 2010, October 29, 2011, February 6, 2015, March 8, 2016, and May 9, 2018. I did also watch Jupiter in retrograde motion in August and September of 1987. Based on the times I have measured for its retrograde loop, the best I can say from my notes is that that opposition was within 60 days of September 29. In the first edition of this book I included that point, but I did give it an uncertainty of ± 60 days. I now have enough photographic measurements to drop the early value from the fit. An Excel plot of the dates of the six recent measurements against their number has a slope of 398.7 ± 0.6 day, so that is my current best value.

I do have more data on Jupiter than on Mars. Take a look at Figs. 15.2 and 15.3 in Part II. The six oppositions of Jupiter that I have measured are concentrated around either its aphelion point or its perihelion. The two intervals between the points near aphelion are 402.7 and 402.6 days, clearly longer than the long-term average. The intervals between the points near perihelion are 396.0 and (2 ×) 395.8 days. The long-term average of 398.7 is about half-way between these.

For Saturn I can try the same approach. I observed its retrograde motion accurately in the springs and summers of 2010, 2011, 2015, 2016, 2018, and 2020. Unfortunately, the earliest, approximate, observation I have was only four years earlier than the first of these. I made a number of sketches of Saturn as it was in retrograde motion, passing by M44, in January and February of 2006. At that time, I didn't try to find the exact moment of opposition, but I can now analyze my sketches much as I analyze star field photographs, although with a lower precision, of course. I find now that opposition occurred on January 27, 2006, DJD number 38743 ± 1. Saturn's ecliptic longitude was 127.5° ± 0.5°. The interval to the most recent measurement is 5288.5 days and this must correspond to 14 synodic periods, since I know from the heliacal setting fit that the period is about 379 days. 5288.5 divided by 14 gives 377.8 days. Based on the 1 day uncertainty in the earlier observation, the purely statistical uncertainty is 0.1. If I fit a linear trend line to all seven observations, the uncertainty is doubled to 0.2.

Uranus was something of an afterthought in this series of observations. During the Fall and Winter of 2010 I took a series of star field photographs centered on Jupiter. It just happened that Uranus was within a few degrees of Jupiter during that time, so I was able to follow its retrograde motions and to locate the time of its opposition with no extra effort. I then remembered that I had taken a star field photograph of Uranus in the summer of 1980, and noted in my log book that it was close to opposition. I based that on the time of night that it passed through my meridian. By comparing that photograph with a star atlas, I could find the right ascension and declination of Uranus at that time. Unfortunately, when I look at the photograph more critically, thirty years on, I can see that Uranus was clearly past opposition. I came up with two ways to estimate how much past opposition it was and, fortunately, they agreed pretty well. One was to follow Uranus after the opposition of 2010 and decide when its position corresponded as closely as possible with the photograph from 1980. This was less accurate than it might have been because the 2010 opposition was much later in the year and the slope and height of the ecliptic were quite different. However, I did my best. The second method was to ignore the slow movement of Uranus relative to the stars, and use the idea that the stars rise earlier by four minutes every day to figure out when the stars around Uranus would indeed have crossed my meridian at local midnight. The two methods both indicated that I had taken the photograph about eight weeks after Uranus was at opposition. This puts the date of opposition as the middle of May, 1980, say May 15 ± 15 days. May 15 1980 was DJD 29355 (at the middle of the day). The opposition of 2010 was on 9/21/2010, at 17.05 hours Universal time, DJD 40441.29. The interval between them is 11086 ± 15 days.. (A painless way to get these intervals without having to sweat out the number of leap years is to calculate Julian day numbers for each date and to take the difference. Duffett-Smith's book has a Julian day subroutine and I have made very heavy use of it.) I can guess that the synodic period for such a distant planet is just over a year, so the Earth must have made thirty passes by Uranus in this time. One thirtieth of the interval is 369.5 ± 0.5 days. I have since tracked oppositions in 2011, 2016, and 2018 and an Excel graph of all five points reduces the uncertainty to ± 0.4 days. It may seem that I have produced a very precise answer from a rather shaky basis, but it is the thirty-year time interval that makes it work. Also, if I try to use this number to calculate the sidereal period of Uranus, what matters is the difference between this number and the Earth's sidereal year. This difference is 4.25 ± 0.4 days, which has almost a ten percent uncertainty. So from that viewpoint there is not much actual information in the value. I need to wait a few more years and try again!

Both Saturn and Uranus have sidereal periods longer than the time for which I have been making measurements, and they have both been moving near their aphelion points. I have no chance of accumulating data comparable with that for Jupiter!

Chapter 6: Sidereal periods of the inferior planets

The sidereal period of a planet is the time that it takes to perform one orbit of the Sun, measured relative to the stars. Extracting this number from a set of Earth-based observations does require some analysis. As I mentioned in connection with the synodic periods, I did manage to follow Venus rather accurately through its maximum eastern and western elongations in the Fall of 2010 and early Spring of 2011, and I finally realized that I had enough information there to calculate Venus's sidereal period. To do this I had to measure the sidereal coordinates, the right ascension and declination, as well as the elongation from the Sun. Actually, in my thinking about the orbits of the planets I usually picture them all as lying exactly in the plane of the Earth's orbit. The position coordinate that gets closest to this is the ecliptic longitude. The ecliptic latitudes of the planets are never more than a few degrees positive or negative. So I used one of Duffett-Smith's subroutines to convert the right ascensions and declinations into ecliptic longitudes. Even with this simplification, I don't think this is an obvious calculation. I wrestled with it for some time and finally got inspiration from a re-reading of Book Five of Copernicus's monumental "On the Revolutions of Heavenly Spheres"! I don't mean that he explains how to do the calculation there, but that he solved similar but more difficult problems so that I felt I should be able to solve this one. So bear with me on this. What I need to find is how far around the Sun Venus has moved between the eastern and western elongations. The first diagram shows the greatest eastern elongation of 8/19/2010. The two circles represent the orbits of Venus and the Earth. I chose arbitrarily to put the zero of ecliptic longitude pointing in the positive-x direction, off to the right side of the diagram. I drew the position of the Earth in its orbit based on the value of the ecliptic longitude of the Sun, which was 146°. I haven't marked this number on the figure because I don't actually make any use of it and I wanted to keep the figure as simple as possible, but it is the angle between the line going horizontally from the Earth and the line going from the Earth to the Sun. The most important number is the ecliptic longitude of Venus, which was 192.18°. The standard mathematical convention is to take the anticlockwise direction as positive, so the line joining the Earth to Venus is rotated 192.18° anticlockwise from the positive-x direction. Because the figure is drawn to scale, this line just touches the circle representing the orbit of Venus, and this locates the position of Venus in the diagram. What I am trying to get out of this is the direction of the line going from the Sun to Venus. This is the heliocentric longitude of Venus. This line is a radius of the circle and must be perpendicular to the tangent line I have drawn. The angle of Venus relative to the Sun is therefore 282.18°. It is useful to notice that this is just the ecliptic longitude of Venus plus 90°.

I could have measured it in the clockwise sense from the positive-x direction, but that is conventionally the negative direction, so the angle would be -77.82°.

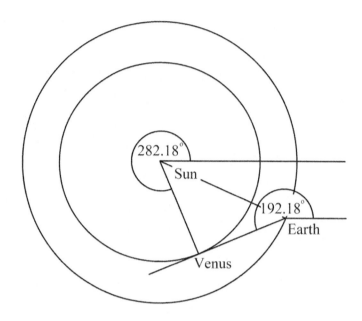

Figure 6.1 Greatest Eastern elongation of Venus in 2010

Now look at the similar diagram for the greatest western elongation. This took place on 1/7/2011, 17 hrs Universal time, when the ecliptic longitude of the Sun was 287°. This locates the position of the Earth on the diagram. I measured the ecliptic longitude of Venus as 240.23° and this allows me to locate Venus on the diagram, where the tangent line touches its circular orbit, and to draw the line connecting the Sun to Venus. This line, the heliocentric longitude of Venus, makes an angle of 150.23° with the positive-x axis. Notice that this is just 90° *less* than Venus's ecliptic longitude. The time interval between the two diagrams is 142.7 days. How far has Venus moved, in an anticlockwise direction? The answer is (77.82° + 150.23°) = 228.05°. The sidereal period is the time for it to turn through 360° and this is

142.7 days × (360°/228.05°) = 225.3 days.

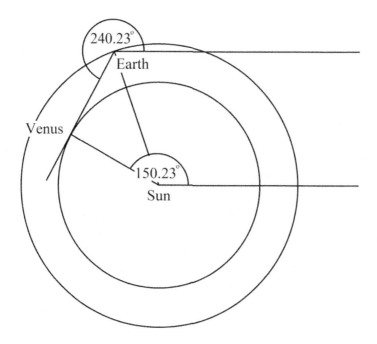

Figure 6.2 Greatest Western elongation of Venus in 2011

Actually the uncertainty in the time interval is about one day, so the decimal fraction is not significant. My value for the period is 225 ± 1 days. There may also be some uncertainty because Venus's orbit is not exactly a circle.

When I had found my way through this calculation, I realized that the usefulness of the maximum elongations was that I could find the heliocentric longitudes at those positions. Another situation where I can find the heliocentric longitude of the planet, actually rather more easily, is the inferior conjunction itself. At that time, the planet is in the same direction as the Sun, when viewed from the Earth, and its heliocentric longitude is just the ecliptic longitude of the Sun plus 180°. From my estimate of the time of the transit of Venus on 6/8/2004, the ecliptic longitude of the Sun was 77.87° and therefore the heliocentric longitude of Venus was this plus 180°, which is 257.87°. I could combine this with one of the maximum elongations of 2010 and 2011, but there is a substantial advantage to making the interval as long as possible. I therefore made a new set of measurements of the maximum Eastern elongation in the Spring of 2015. I found that the maximum elongation occurred at 18:00 hours Universal time, on 6/6/2015. This is DJD 42160.25. The ecliptic longitude of Venus was 120.98°. Since this was an Eastern elongation, the heliocentric longitude of the planet was this value *plus* 90°. or 210.98°. The time interval from the 2004 transit to this point is 4015.40 days. Since I know from the previous calculation that the sidereal period is about 225 days, I can find that Venus has completed almost 18 revolutions about the Sun. This means that any effect of ellipticity of the orbit will be very small. The total angle turned through by Venus is actually

$$18 \times 360 + 210.98 - 257.87 = 6433.11°.$$

There is a small uncertainty in this because the 90° difference between the ecliptic and heliocentric longitudes of the planet at maximum elongation is exact only for a circular orbit. However I expect the uncertainty for Venus, which has an orbit with very low eccentricity, to be less than 1°. If Venus has turned through 6433.11° in 4015.40 days, the sidereal period, which is the time for it to turn through 360°, comes out at 224.70 ± 0.03 days, or 0.61518 ± 0.00008 Earth years. I have made several more measurements of both the maximum Eastern and maximum Western elongations of Venus, which are all described in Part II of the book, and I can use them to get what should be even more accurate values. The graph is shown here.

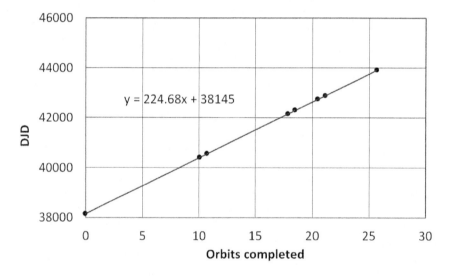

Figure 6.3 Venus' sidereal period

What is plotted along the x-axis is actually the number of orbits completed. For each data point I used the value of 224.7 days to figure out how many orbits had been completed since the first point and added on the fraction of an orbit corresponding to the heliocentric longitude.

The slope this time is 224.68 ± 0.03 days or 0.61514 years. However, this is the first time in the book that I have reached a useful limit. The Observer's Handbook lists the orbit parameters for each of the planets, giving values for

January and July each year. This is necessary because the orbits are not exactly ellipses and the numbers represent the best fit to an elliptical orbit at that time. For Venus in 2020 the numbers for the sidereal period are 0.6152 years in January and 0.6021 years in July. So perhaps my extra decimal places are not meaningful. (In 2021 the Handbook listed long-term averages rather than osculating values.)

I can make a somewhat similar calculation of the sidereal period of Mercury. In the first edition of this book I relied mostly on relatively imprecise measurements that I made before I started using the star-field approach. I did manage to extract an approximate value of 88.2 days for the sidereal period. I have now managed to follow Mercury through two Eastern elongations, although I can usually see it only on a few days during each elongation. For the Western elongations I relied on the handbook to tell me when the maximum elongation occurred. So I have only a single value for them. I do have good values for the longitude of Mercury during the two transits that I mentioned in Chapter 4, since the ecliptic longitudes are the same as for the Sun and the heliocentric longitudes are this plus 180°. I can plot the six points on the same type of plot that I used for Venus, using the old value of 88.2 days to figure out how many orbits had been completed between observations. Here is the graph.

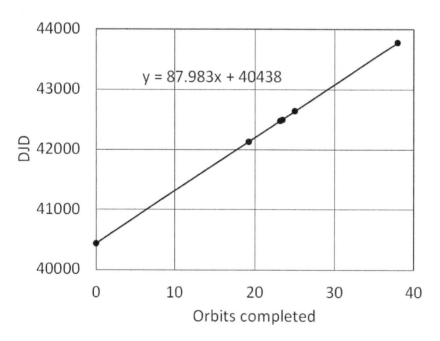

Figure 6.4 Mercury's sidereal period

Time measurements

The first number in the equation on the graph is the sidereal period. The uncertainty is 0.09 days so I should probably write the answer as 88.0 ± 0.1 days, or 0.2409 years.

This is, in fact, a fairly accurate result. 0.09 is the statistical error arising from the quality of the fit. There is potentially another contribution to the uncertainty because in analyzing the four maximum elongations I have relied on the assumption of circular orbits in arriving at the idea that the hour angle of Mercury is equal to its ecliptic longitude plus or minus 90°. The orbit of Mercury is one of the most elliptical of all the planets and the angle between the line from the Sun to the Mercury and the tangent to the orbit can differ from 90° by as much as 12°, depending where Mercury is in its orbit. I have no way of determining this for myself. Based on the distances from the Sun described in Chapter 13 the two Western elongations were very close to perihelion, so the 90° assumption should be good but I don't know about the Eastern elongations. I also have the feeling that I have relied on too much input from the handbook or from Duffett-Smith's programs to feel completely satisfied with the result. I shall keep trying.

Chapter 7: Sidereal periods of the superior planets

For the outer planets the measurement is, I think, conceptually easier. It is now quite easy to see or photograph the planets against a backdrop of stars, but there is again the difficulty that we are viewing the planets from a moving platform. Most of the time, a large part of the apparent motion of the planets arises from the motion of the Earth. The only time that this parallax effect is eliminated is when the planet is in exact opposition to the Sun. At that time you know that you are looking at the planet in the same direction as it would be seen from the Sun. Most of my analysis for the inferior planets was to extract their heliocentric longitudes from the Earth-based measurements. But for the superior planets at the point of opposition, their heliocentric longitudes are the same as their ecliptic longitudes. Especially for Saturn and Uranus, the sidereal period is quite long, so my observations cover only a fraction of an orbit. I make the simplest assumption that they are moving at constant angular speeds to find the time for a complete orbit. For Mars and Jupiter, my observations extend over five complete orbits plus a fraction in the case of Mars and over two complete periods in the case of Jupiter. There is no assumption of uniform motion for the complete orbits, so the calculated periods should be more accurate.

From my observations of retrograde motion in 2007-2008, Mars was at opposition 12/25/07, DJD 39439.5 with right ascension 6 hr 9.0 min, or 92.248° expressed in degrees, and declination 26.745°. The most recent opposition I have measured was on 7/27/2018, DJD 43306.773. Just as in the previous chapter, I decided that the best number to look at for the position was the ecliptic longitude, which measures the distance that the planet has moved around the ecliptic. The ecliptic longitude is an Earth centered measurement. A better number is the heliocentric, or Sun-centered, longitude. Fortunately, as I just pointed out, at opposition the two values are the same. There is a small Basic program in Peter Duffett-Smith's book that does the calculation. The values for the ecliptic longitude for the two oppositions are 92.470° and 304.200° respectively. The difference is 212.730°. There is a trap here because Mars has actually performed five complete circuits plus this much in the time interval of 3867.6 days, so the sidereal period is

$$3867.6 \times 360 / (5 \times 360 + 212.73) = 691.76 \text{ days} = 1.895 \text{ years.}$$

In calculating the synodic period of Mars I used an approximate observation of an opposition in 1990, taken from my old log books. Unfortunately the naked-eye observations did not give me an accurate value for the ecliptic longitude and hence the heliocentric longitude. So I am

41

Time measurements

limited to the relatively short period that I have been using the photographic methods. If I fit a straight line to the recent measurements the period is 1.897. If I include the earlier value it becomes 1.885, so I am calling my best estimate 1.89 ± 0.01 yrs.

For Jupiter, the oppositions that I have to work with are six fairly carefully observed ones from August 2009, September 2010, October 2011, February 2015, March 2016, and May 2018, and a measurement from a less complete set of an opposition in September 1987. The long time interval between the first, approximate, measurement and the more recent ones outweighs the higher precision of the closely spaced ones. My best bet is to use the early measurement and one of the more recent of the high precision ones. Using the most recent measurements also has the advantage that Jupiter has actually completed two revolutions since 1987, so that there is very little error in assuming that it has moved at constant angular speed for the small extra part-revolution.

I estimated Jupiter at the earlier opposition to have right ascension 25.5° and declination 8.75°. The ecliptic longitude corresponding to this is 26.8°. At the opposition of February 6, 2015 the ecliptic longitude was 137.6°. The difference is 110.8° and the difference in Julian day numbers is 9992.8. This time, Jupiter has actually performed two complete circuits plus this much in the time interval of 9992.8 days, so the sidereal period is 9992.8 × 360/(2×360 + 110.8) = 4330 ± 26 days = 11.86 ± 0.07 years. The uncertainty comes from the 60 day uncertainty in the earlier opposition. There are extra uncertainties arising from the use of the ecliptic longitude and the assumption of uniform motion, but they should both be small. If I carry out a fit including all my recent measurements the value changes slightly to 11.91 ± 0.06 years.

In the case of Saturn, I can use the same observations that I used to find the synodic period. I also now have an accurately observed opposition on July 20, 2020. I measured the ecliptic longitude of Saturn to be 298.66° (I go into detail on how I measure these values in the chapter on distances). The other opposition was less well observed. The values are based on hand-drawn sketches. This sounds a bit hairy, but Saturn was moving across the part of the constellation Cancer that contains M44. This is a fairly compact and recognizable region so I think my numbers are quite accurate. The time interval between the two oppositions is 5288.5 days. The values for the ecliptic longitude are 127.5° on 2/3/2006 and 298.66° on 7/20/2020. This means that Saturn had moved 171.2° round the ecliptic in 5288.5 days, so it would get completely round in 5288.5° × (360/171.2) = 11,123 days, or 30.5 years. The purely statistical uncertainty based on these two points is only about 0.1 years, but I am extrapolating quite a short arc into a complete orbit. If I do a least-squares fit to all seven of the oppositions that I have observed, the best value increases slightly to 30.6 ± 0.5 yrs, but to be cautious I shall bump the uncertainty up to 1.0 yr.

For Uranus, I have the same fortuitous observations that I used for the synodic period. In fact, they work a bit better in this case. I did measure very precisely the position of Uranus at its opposition of 2010. The time of the opposition was DJD 40441.29 and the ecliptic longitude of Uranus was 358.6°. From the photograph I took in 1980, I could read off the right ascension as 15 hr 18 min, and the declination as -18.0°. At that date, these correspond to an ecliptic longitude of 231.2°. From my measurements of the retrograde loop in 2010, I know that in the previous eight weeks, the ecliptic longitude of Uranus would have changed by less than 2°, so the ecliptic longitude at opposition would have been close to 233°, so Uranus has moved through (359 – 233) = 126°. The time interval between the two oppositions was 11086 ± 15 days, so the sidereal period is

$$(11086 \pm 15 \text{ days}) \times (360°/126°) = 31674 \pm 44 \text{ days} = 86.7 \pm 0.1 \text{ years.}$$

As usual, the uncertainty is based only on the uncertainty in my measured values. There is no allowance for the assumption of uniform circular motion. Since my measurements cover only about one third of an orbit for Uranus, this is certainly the biggest source of error, but I have no way of estimating it based solely on my observations. The discrepancy from the handbook value is in fact much larger than the 0.1 years statistical uncertainty. The same is true for the results for Saturn, and I did bump up my estimate of the uncertainty in that case. In fact, a fit to all my values for Uranus gives a sidereal period of 87.6 ± 0.4 yrs, and again out of caution I shall increase that to ± 1.0 yrs to allow for the sad fact that all of my observations are at the aphelion end of the orbit.

You might question how I knew that I had to put in extra multiples of 360° for Mars and Jupiter when I didn't do that for Saturn or Uranus. In chapter 14 I shall describe another way to get at these values that turns out to be not as accurate but which does give me unambiguous ball-park values. I used those values to guide these calculations. Copernicus had worse problems to deal with.

Part II: Distance Measurements

Chapter 8: One to get started - Latitude and Longitude

Measuring your latitude and longitude may not be exactly astronomy, but the reduction of the measurement does involve some astronomical ideas, and both the method and the equipment come in useful in other projects.

The classic method of finding your position on the Earth involves using a sextant to find the elevation of the Sun at noon and the exact time at which this occurs. You can spend quite a lot of money on a sextant if you so choose. I chose not to and bought an inexpensive plastic sextant manufactured by Davis Instruments, http://www.davis.com/, their Mark 3 model. I bought mine in 1987 but it was still available in 2020. In terms of entertainment per dollar it is great but it has three limitations. The first is that it does not contain a telescope; you are looking directly, naked eye, at the targets. This limits the precision that you can set it with and, if you use it to look at stars, you can see only those that are naked-eye visible. The second limitation is that the scale is, like the rest of the instrument, made of plastic, with a simple vernier for more accurate readings. I have trouble setting and reading the vernier to better than two or three minutes of arc. I told myself that this is the sort of accuracy that Tycho Brahe could get from his instruments, and what was good enough for him is surely good enough for me. The third limitation is that the main scale stops at 100 degrees. As a result I can use the false horizon method (see below) only when the elevation of the Sun at noon is less than 50 degrees, which means for my location that it is not do-able between the Spring and Fall equinoxes.

If you do have a good horizon, at the shore for example, the method of operation is to set the pointer to zero degrees, put appropriate filters in the light path and look up at the Sun. Then gradually move the pointer away from you while lowering the angle of the sextant to keep the reflected image of the Sun in view. Adjust the pointer until the lower edge of the image of the Sun makes contact with the horizon. Rock the sextant from side to side to make sure the image is at its lowest point when it touches the horizon. This last step is very important. Perhaps because the motion of the Mark 3 has some friction, it is tempting to tilt the sextant slightly to make a small adjustment to the height of the Sun on the horizon. It took me quite a lot of practice to break this habit. When you have the image sitting properly on the horizon, take the scale reading, making use of the vernier. There are two corrections, one is to add on half the angular size of the Sun, approximately

Distance measurements

16'; the other is a correction, tabulated in the instructions, if your position is above the height of the horizon.

If you don't have a good horizon, the recommended method is to use a false horizon. This consists of a horizontal reflecting surface which could be just a bowl of water, although there are more sophisticated versions. This time you adjust the sextant until the image of the Sun seen through the sextant and the image reflected in the bowl of water are touching and then divide the answer by two. This is the step where the 100° maximum on the scale reading limits the times of year when the measurement can be made. You still need to add on half the angular size of the Sun but the correction for the observer's height above the horizon is not needed in this case.

In either case, start taking readings at least half an hour before local noon and continue them at regular intervals until half an hour after noon. You need to pick out the maximum value of the reading to get the latitude and the time when it occurs to get the longitude. It may be overkill, but I like to put the points into Excel and fit a second degree curve. From that you can get the maximum scale reading and the time.

The processing of the results contains several details. If you find yourself doing this more than a very few times, you might choose to develop a spreadsheet to automate the process. To find the latitude, take the maximum elevation of the Sun and subtract it from 90°. Add to this the declination of the Sun. You can get this last number from an almanac such as the "Observer's Handbook", published each year by the Royal Astronomical Society of Canada, or you can calculate it using some of Peter Duffett-Smith's subroutines The instruction booklet that came with my sextant included a table of values, but be aware that it goes slightly out of date. To get the latitude good to one or two minutes you have to be precise in every step – alignment, taking the readings, and processing the result.

If you have the "Observer's Handbook", the easiest way to get the longitude is to look up the time of local noon at the Greenwich meridian. This has the so-called equation of time built into it. Convert that to your time zone, or convert your time to the Greenwich time, and the difference between that value and the time you found for the maximum elevation of the Sun is your longitude expressed as a time. Multiply by 15° for each hour of time to put it in degrees. The instruction booklet warns that it is difficult to be more accurate than the nearest 10 minutes of arc. To achieve that, you again have to pay attention to every detail, including the timing. An error of one minute in the time when the maximum occurs corresponds to an error of a quarter of a degree, or fifteen minutes of arc, in the longitude.

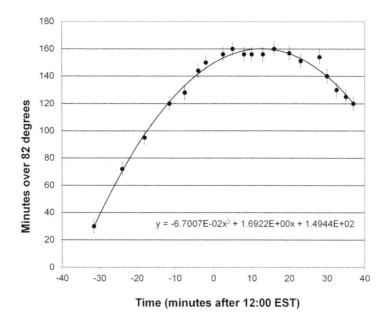

Figure 8.1 Altitude of Sun on March 1, 2011

As you can imagine, I have made this particular measurement many times, using a variety of ways to overcome the lack of a good horizon at my home. Figure 8.1 shows the sort of plot I get. This is data taken on March 1, 2011. I used a black photographic tray containing water as a false horizon and waited for as windless a day as possible. To avoid working with big numbers, I have plotted the times as minutes before or after 12:00 noon EST, and I have subtracted 82° from all the angle readings and plotted the difference in arc minutes. There is some scatter of the points. I expect that fitting the line will smooth out the random errors. Either by reading the graph carefully or by calculation from the fitting parameters, I find that the maximum of the curve is 160 arc minutes over 82°, which is 84° 40', and it occurs at time 12.6 minutes after 12:00 noon.

The calculation of the latitude is then:

Maximum angle (false horizon reading)	= 84° 40'
Maximum elevation of Sun = half of this	= 42° 20'
Lower limb correction	= 16'
Corrected elevation	= 42° 36'
90° minus this	= 47° 24'
Declination of Sun on 3/1/2011	= -7° 32'
Latitude = sum of last two	= 39° 52'

47

Distance measurements

I estimate the uncertainty in an individual reading to be 4 or 5 minutes, and the uncertainty in the fitted maximum to be ±2 minutes.

For the longitude:

Time of maximum elevation	= 12:12:36 pm EST
	= 17:12:36 UT
Local noon at Greenwich	= 12:12:22 UT
Difference	= 5hr 0m 14s
	= 5.00389 hr
Times 15°/hr = longitude (West)	= 75.0583°
	= 75° 4'

I wouldn't bet any money on the time of the maximum being more accurate than ± 1 minute (of time) and, as I mentioned earlier, this corresponds to an uncertainty in the longitude of ±15 minutes (of arc).

The declination of the Sun came from the Observer's Handbook. It lists values for 0 hr Universal Time on February 26th and March 2nd. I interpolated to 17:00 UT on March 1st. The local noon at Greenwich also comes from the tables and I interpolated to March 1st.

So my best values are 39° 52' ± 2' North, for the latitude, and 75° 4' ± 15' West for the longitude. Google Earth gives 39° 53' 20" N and 74° 56' 23" W. I don't think I can do any better with my plastic sextant.

Of course, the astronomers of two thousand years ago didn't have even a plastic sextant. They presumably used some variation of measuring the length of the shadow cast by a vertical stick. I have tried some implementations of this. I got good results using a plumb bob on a line instead of a stick. For a horizontal scale I placed a yardstick, with a spirit level sitting on it, at the tip of the plumb bob. The shadow of the string becomes invisible after a few inches, so I clamped a larger marker on the string about five feet above the ground. I estimated that I could measure the altitude of the Sun to about a tenth of a degree accuracy. That is not quite as accurate as the false-horizon measurements, but is competitive in the summer when I can't use the false horizon method. Once you have the altitude of the Sun, the reduction of the data is the same as for the sextant results.

There are several ways to extract more astronomical numbers from these measurements. If you measure the highest altitude of the Sun at regular intervals through the year, you will find that it reaches a maximum during the summer and a minimum in the winter. Actually, the maximum and minimum occur at the summer and winter solstices, but by making regular measurements you can locate them for yourself without the extra reliance on an almanac. Whatever latitude you live at, the difference between the summer maximum and the winter minimum gives you the latitude difference between the tropics of Cancer and Capricorn, and half of this number is the angle that the Earth's rotation axis makes with the normal to its orbit round the Sun, or, equivalently, the angle between the ecliptic and the equator. My own measurement of this is 23° 26', which agrees with the Handbook value to the nearest minute. There was an element of good luck in this. Each of my values of the altitude of the Sun (at the two solstices) was a few minutes high, and this canceled in the difference. So I estimate the uncertainty in my value to be ±5'.

If you keep up the measurement of the maximum altitude of the Sun for several years, you can extract a value for its repeat time, and this is the tropical year. I have been making observations for this purpose rather intermittently since 1987. The best number I have pulled out so far is 365.245 ± 0.01 days. I am a little disappointed that I can't pin down the uncertainty better. The difference between this value and the length of the sidereal year, which I measured in chapter 3, is caused by the precession of the Earth's rotation axis, and this seemed to me to be my best route to measuring that number. However, at the moment, my value for the difference between the two years is also equal to my estimate of the uncertainties in the two numbers, so I really can't say anything about it.

Chapter 9: Radius of Earth

The classical way to find the radius of the Earth is to measure the latitudes of two points a known distance north and south of each other. Eratosthenes of Alexandria is said to have used this method around 200 BC. I wrote about measuring the latitude in my backyard to be 39° 52' in the previous project. A few years ago I was visiting West Cape May, at the southern tip of New Jersey, and I realized that I was almost exactly due south of my home. I couldn't resist that nudge. I took my plastic sextant down to West Cape May and measured the latitude as 38° 58'. My latitude measurements have uncertainties of about 2', so the difference of two latitudes has about a 3' uncertainty. (If the errors are random, the way to combine the two uncertainties is to square them, add them together, and take the square root. To one significant figure this gives 3'.) Since the difference of latitudes is only about 1°, or actually 54', I am limiting myself to a 3 parts in 54, or 6%, accuracy. I could improve that by choosing two points that were further apart, but that would increase the difficulty of actually finding the distance between them, and I also like the idea of having one point in my own back yard. So I accepted West Cape May as a compromise.

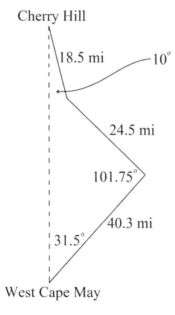

Figure 9.1 My route from Cherry Hill to West Cape May

The next step is to find the straight line distance between the two locations. If you look at a map of New Jersey you will see that the estuary of the Delaware River sticks into a straight line path from Cherry Hill to Cape

Distance measurements

May so I am not able to drive from one to the other in anything like a straight line. I divided my usual route into three stages that I could plausible approximate as straight lines. The two longer sections are expressways, which tend to be fairly straight. I drove that route and measured the lengths of the three stages, and I estimated the angles between them from a map of the state. That gave me the picture shown in Figure 9.1. I could now calculate the vertical component of the trip as 69.4 miles. (This is explained more fully as an example in Appendix A.) The best measurement I could make of the distance on the state map was 315 ±1 mm, and multiplying this by the scale thoughtfully printed on the map gave me 66.1 ± 0.2 miles. My personal answer is 5% high, which seems plausible.

There is another approach that might occur to you. If I open up Google Earth in my computer (in 2016), click on the ruler, and then click on the two end points, the distance between them is displayed. I am pretty sure that Google Earth calculates the distance from the latitude and longitude of the two points plus a model of the shape of the Earth. It is putting into the calculation just the quantity I am trying to find. So I regard that as cheating. I am hoping that my state map is based to some extent on old-fashioned surveying methods. (I actually get a distance of 66.3 miles if I do use Google Earth.)

The distance r between the two points and the difference in latitude between them, θ in radians, are now related to the radius of the Earth by the usual formula for an arc length (see main Appendix A if this is not familiar).
$$r = R\theta \quad (\theta \text{ in radians})$$
$$\therefore R = \frac{r}{\theta} = \frac{66.1}{0.015708 \pm 0.0009} = 4208 \pm 234 \text{ miles.}$$

The uncertainty is a bit high, but, as I explained earlier, I knew it was going to come out that way before I started the measurement. The usual scientific practice is to round the numbers to eliminate figures that are not meaningful. So I would round my result to 4200 miles. The uncertainty falls in an awkward range; if I round it to 200 miles I am being optimistic, but if I round it up to 300 miles I am being pessimistic. In the list of values in the introduction I did take the cautious approach, and wrote is as 4200 ± 300 miles or 6750 ± 500 kilometers. This calculation ignores the idea that the Earth is not exactly spherical, but I shall not try to tackle that correction until I can reduce the uncertainty in my value.

You might enjoy taking a look at the related measurement described in the book by Sydney G. Brewer referenced earlier and in the bibliography. He describes how he measured the distance between two points 44 km apart with less than 0.5% error, starting from a one-foot steel rule and using a bicycle with a calibrated rear wheel. Both Mr. Brewer and Dr. Clark, in their books referred to in the introduction, used a different measurement to find the size of the Earth. They measured the time difference between observations of the same astronomical event seen from two locations at a known distance apart and lying due east and west of each other. I chose to go the historical route, but you might prefer theirs.

Chapter 10: Calibrating your camera

In the preface, I mentioned that I have been using digital photographs more frequently, actually since digital cameras made their appearance. I have used photographs of the Sun, Moon and planets in the measurement of their angular sizes and I have used photographs of the fields of stars around a planet as a step in the determination of their coordinates – right ascension and declination. I use a Nikon D80 camera, either mounted at the prime focus of a small refracting telescope with 910 mm focal length, or with a 50 mm lens or a telephoto zoom lens at its maximum focal length of 300 mm. The camera is getting a bit out-of-date, but there is some advantage to using the same equipment for a whole series of measurements so I am sticking with it for now. I want to step through the calculation of an angle from such a photograph, and then to think about the need for calibration. If your trigonometry is rusty, look at Appendix A. Appendix B contains the thin lens formulas.

Suppose that a lens focuses an image on a screen a distance i away (i stands for image distance).

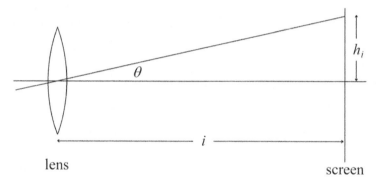

lens   screen

Figure 10.1 Geometry of image formation

The angle θ is related to i and to the image height h_i by

$$\tan(\theta) = \frac{h_i}{i}.$$

In a typical astronomical application of this, the base line of the triangle would be very close to the focal length, f, of the lens. Many image processing programs will allow you to read off the positions of points in the image measured as numbers of pixels horizontally and vertically from one corner. Suppose you have an image of the Moon stretching from the center of the image to a point n pixels above center. From the specifications for your camera, you can find the total size of the sensor and the number of pixels in each direction. Divide one of these by the other to get the size of one pixel. Let's call it d. Then the distance h_i would be nd and the equation for θ would become

$$\theta = \tan^{-1}\left(\frac{nd}{f}\right).$$

Actually, you are more likely to have tried to center the Moon in the image. The proper procedure then is to split the calculation into two parts. Find the number of pixels from the center of the sensor to the top of the image, and from the center of the sensor to the bottom of the image, calculate the angle corresponding to each distance (using a scientific calculator that has an inverse tangent function) and add them together.

In many cases, there is a big short cut, based on the properties of small angles. If the focal length is much larger than the sensor size, the angle θ might be a few degrees or less. You can then use the property that the tangent of a small angle is almost equal to the angle itself, provided the angle is measured in radians. For example, suppose θ is 5°. To convert that into radians, multiply it by π and divide by 180°. This gives 0.08726 radians. The tangent of 5° is 0.08748. The difference is 0.2%. For even smaller angles the approximation gets better and better. The angular size of the Moon is about 0.5° and the tangent differs from the angle in radians only in the fifth significant figure or about 0.003%. If that is adequate, then the angular size in the previous example would be just nd/f radians, and you convert that back to degrees by multiplying by 180°/π. Whether this is the case or not, the value of θ involves the ratio d/f.

So far, so good. In my case, the Nikon D80 has a sensor that is 23.6 mm wide by 15.8 mm high, and the number of pixels is 3872 × 2592. The distance d is then either 0.0060950 mm or 0.0060957 mm. This is an insignificant difference, so I can take an average and call it 0.0060953 mm. The focal length of the objective in the refracting telescope that I used for the prime focus photographs is 910 mm. So the ratio d/f is 6.6981 × 10⁻⁶. Well, maybe. The difficulty is that the values for d and f are nominal ones. I don't know how consistent the d value is in the sensor but I do know that the focal lengths of lenses can vary by several percent from their nominal values. If I am trying to find the angular size of the Sun or Moon, I would like to do better than a few percent, and to do that I need to calibrate my equipment.

The procedure I used to calibrate the telescope/D80 combination was to mount it in daylight in my back yard, pointing at a yardstick some distance away. From a measurement of the distance between two of the markings on the yardstick on the digital image, I can find a calibration. In fact, the front of the telescope was 185 ft 6 in from the target. This converts to 56,540 mm. (This is an awkwardly small unit but all of the focal lengths are given in mm.). I found that two markings on the yardstick that were 28 in, or 711.2 mm, apart were separated in the image by 1936 pixels.

There is a correction needed because the yardstick is not infinitely far from the telescope; my back yard just isn't that big! If the object is a distance o from the lens, and has a height h_o, the complete expression for d/f is

$$\frac{d}{f} = \frac{h_o}{n(o-f)}.$$

I have put the derivation of this into an appendix at the end of the chapter. The result looks slightly circular because there is an f on the right side and I have said that I don't know the exact value of f. However, the f on the right side of the equation is a small correction to the much larger value of o, so using the formal value for f will be adequate. With the numbers that I found in my case

$$\frac{d}{f} = \frac{711.2}{1936(56,540-910)} = 6.6035 \times 10^{-6}.$$

This is about 1.4% lower than the nominal value. I estimate the uncertainty in this is 0.2%. To get results to 1% accuracy or better, you must make this calibration. If all of the discrepancy is caused by a variation in the focal length of the telescope objective, its actual value is 923 mm.

I have also taken photographs of the Moon using a zoom lens set to its maximum focal length of 300 mm. I used a similar procedure to calibrate it. The tolerance on actual values of the limiting focal lengths for a zoom lens is quite large, so you really need to do the calibration in this case. For the 50 mm lens that I use for star-field photographs, I actually used a star-field image to find the d/f value. I identified several stars on the photograph, looked up their right ascension and declination coordinates and found the angles between them. Then I adjusted the d/f value until my calculated separations agreed.

Distance measurements

Appendix: Calibration formula

Two formulas from Appendix B are

$$\frac{1}{o} + \frac{1}{i} = \frac{1}{f}$$

and

$$\frac{h_i}{h_o} = (-)\frac{i}{o}.$$

o is the object distance, i is the image distance and f is the focal length. I can solve the first equation for the image distance i as

$$i = \frac{of}{o - f}$$

and use that in the second equation solved for h_i

$$h_i = \frac{ih_o}{o} = \frac{ofh_o}{o(o - f)} = \frac{fh_o}{o - f}.$$

The image height is equal to nd, where n is the height measured in pixels and d is the height of one pixel, and so

$$nd = \frac{fh_o}{o - f}$$

$$\therefore \frac{d}{f} = \frac{h_o}{n(o - f)}.$$

This is the formula I need.

Chapter 11: Analyzing Star Field Photographs

In many of the results I have already mentioned, and especially in connection with trying to extract distances inside the solar system, it is almost essential to be able to find the celestial coordinates, the right ascension and declination, of the Moon or planets. Over the years my technique for doing this has improved. I started by looking at the object through binoculars and estimating its angular distance from two or three stars. Knowing the field of view of the binoculars I could do this to maybe a quarter of a degree accuracy. From a star chart I could then estimate the coordinates of the object. An improvement on this was to take a photograph of the field of stars around the object. I standardized on one lens, a 50 mm fixed focal length Nikkor lens, and simply mounted the camera on a fixed tripod. With the ISO set at 800, an exposure of 1 to 5 seconds at f/2.8 usually gives a good result. The next step was to compare the photograph, displayed on the computer monitor, with a good star atlas, usually the SkyAtlas 2000.0, of Will Tirion and Roger W. Sinnott, perched on my knee, and to estimate the point on the star chart that corresponded to the position of the planet in the image. This reduced the amount of scatter, but I still thought I could do better. More importantly, I wanted a procedure that was based entirely on measurements on the digital photograph and did not depend on my judgement at any stage.

What complicates this problem is that the right ascension and declination are coordinates on a spherical surface whereas the photograph is flat. I am sure you have all looked at Star Atlases and seen that the lines of constant right ascension and declination are curved, more in some directions than others. If all the objects of interest are close together on the photograph, this curvature can be ignored. The transformation from pixel coordinates on the photograph to celestial coordinates is then relatively straightforward. I have put the equations in an appendix. It turns out that they contain five quantities that need to be determined in order to convert from one set of coordinates to the other. If you measure the pixel coordinates on a photograph for three stars, and find their right ascensions and declinations from a star catalog or a planetarium computer program, you have six relationships between the two coordinate systems. This over-determines the solution. You could throw one of them away, but it is better to keep them all to preserve the symmetry of the equations. The over-determination then leads to a consistency check so that you can see just how adequate the neglect of the curvature, or the linear approximation, is. You could do the calculations by hand, but if you are doing a lot of them, such as following a planet through its retrograde motion, it is better to automate it. I have set up the solution as an amazingly compact Excel spreadsheet.

But situations will come up where you cannot restrict yourself to a small area of the sky. Sometimes the sky is cloudy and you are picking out the stars you can recognize in the gaps in the clouds. If you are trying to track one of the inferior planets, they are usually close to the horizon in a sky that is still light. The only visible stars may be much higher in the sky. So finally, I developed a computer program that makes no assumptions of linearity in the coordinates, so it is not restricted to a narrow field of view. It converts the pixel coordinates of points on the photograph into values of right ascension and declination to within about 15 seconds of arc accuracy. I originally programmed this in TrueBasic. More recently I have also adapted this calculation into what is still a surprisingly compact spreadsheet. The math is again described in the appendix.

In use, I take a star field photograph and identify on it three stars. It is worth taking some care in the selection of stars. The stars should not lie close to a straight line, but form a well-defined triangle. Generally it works best if I select three stars that are fairly close to the object I am locating, rather than going straight to the brightest stars. It is also best to avoid stars close to the edges of the photograph. Most lenses introduce some distortion, either barrel or pincushion distortion, or coma, near the edges, and the program makes no allowance for that. Finally, take care to locate the center of each star image to the nearest pixel. I occasionally try to estimate the nearest half pixel. As I mention elsewhere, my camera is getting a little elderly, and a new model with more pixels might be slightly easier to use.

I enter into the program or spreadsheet the pixel coordinates of these three stars together with their right ascensions and declinations, which I get from a planetarium program. Then I enter the pixel coordinates of the planet or Moon or whatever I am tracking, and the program more or less immediately gives me the right ascension and declination. I am so pleased with the results that I have been using this method for almost all my recent measurements. In principle, there is a correction to be made to these numbers, to convert values based on my position on the surface of the Earth to numbers that would be measured by an observer at the position of the center of the Earth. It is quite a small correction and is much smaller than the precision of my measurements for the more distant planets. Venus and Mars do approach much closer to the Earth and the correction is getting to be comparable with my precision. I have ignored it so far, but would have to include it if I managed to improve my measurement technique. I do bring this up again later when I talk about measurements of the Moon's position. This correction is essential there.

I go over the math background for the calculations in the end-of-chapter appendix. This is probably the most demanding piece of math in this book, but it really was worth my while to develop it. The Excel spreadsheets for both versions of the calculation can be downloaded from the web page www.GaugingtheSolarSystem.com.

Appendix: Linear and Spherical transformations

Given a photograph of the surrounding star field, many image manipulation programs allow you to find the positions in pixels of the object you are interested in, usually a planet, and of three stars around it, referred to a familiar pair of orthogonal axes, that I shall just call the x- and y-axes. The simplest approach is to suppose that the stars and planet of interest are close enough together in the sky that the curvature of the celestial plane around them can be neglected. Then a second set of orthogonal axes can be used to locate them. The mathematical problem is then to convert the x- and y- positions of their images on the photograph into the corresponding right ascension and declination positions on this second set of axes. All the image processing programs that I have used measure the numbers of pixels counting from the top left hand corner of the photograph. Compared with the usual mathematical usage this is a left-handed set of axes. As it happens, the right ascension and declination also form a left-handed set of axes when you view the sky from the center of the celestial sphere. Their orientation relative to the axes on the photograph depends quite strongly on the latitude and the direction in which the camera was pointing, so I need to have one angle, θ, to measure that. The origin of the (x, y) axis system is on the photograph, but the center of the celestial axes is somewhere out there! I need two numbers to measure the separation of the origins of the two sets of axes. In my spreadsheet I subtract from all of the x's and y's their values at the center point, though that is not necessary in this linear calculation, and this puts the origin of that set of axes at the center point. Let me take this point on the photograph to correspond to right ascension and declination values of R_0 and d_0. (I am using R for right ascension for brevity.) Finally, I need two more parameters to relate the scales of the two axes. The position values on the photograph could just be left as numbers of pixels, but for the non-linear approach that I shall talk about next a less arbitrary choice is necessary. I multiply the numbers of pixels by the size in millimeters of each pixel and divide by the focal length of the lens, also in millimeters. At least near the center of the photograph the values are then the angles, in radians, subtended at the lens. For the celestial coordinates, I convert the hours, minutes, and seconds of the right ascension into an angle in degrees, and also express the declination in degrees. As you can see in any star atlas, the lines of right ascension get squeezed together as the declination increases, so I do need adjustable scale parameters to make the two sets of axes correspond. I shall call them a and b. A side consequence of using these scale parameters is that the final answers are independent of the units you use for x and y. The geometry of the two sets of axes is shown in Figure 11.1 below.

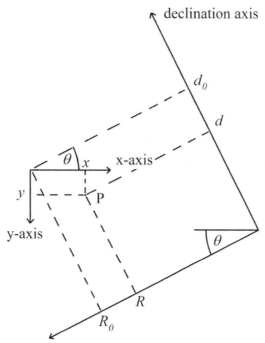

Figure 11.1 Relationship between the two sets of axes

Suppose one of the measured objects is at point P. Its coordinates on the photograph are x and y. They are the projections on the x- and y-axes. As shown, the origin of the (x, y) axis system projects on the right ascension and declination axes at R_0 and d_0. With some trigonometry you can calculate the celestial components of the position of the object as

$$R = R_0 - ax\cos\theta + ay\sin\theta,$$
$$d = d_0 - bx\sin\theta - by\cos\theta.$$

Now suppose that for the three stars surrounding the planet I have measured their positions on the photograph as (x_1, y_1), (x_2, y_2), and (x_3, y_3), and I have found their celestial coordinates from a star atlas or computer program as (R_1, d_1), (R_2, d_2), and (R_3, d_3). I can write out the equation for the right ascension of each star putting in the appropriate values. I have reversed the sides of the equations. The results are

63

Distance measurements

$$R_0 - (a\cos\theta)x_1 + (a\sin\theta)y_1 = R_1,$$
$$R_0 - (a\cos\theta)x_2 + (a\sin\theta)y_2 = R_2,$$
$$R_0 - (a\cos\theta)x_3 + (a\sin\theta)y_3 = R_3.$$

Remember that x_1, y_1, R_1 etc. are just numbers, so I can think of this as a set of three simultaneous equations in the variables R_0 and the combinations $a\cos\theta$ and $a\sin\theta$. While it is possible to solve these equations by direct eliminations, it is well worth developing a more efficient method, especially if you are expecting to do this often. I can rewrite these equations using a matrix notation as

$$\begin{pmatrix} 1 & -x_1 & y_1 \\ 1 & -x_2 & y_2 \\ 1 & -x_3 & y_3 \end{pmatrix} \begin{pmatrix} R_0 \\ a\cos\theta \\ a\sin\theta \end{pmatrix} = \begin{pmatrix} R_1 \\ R_2 \\ R_3 \end{pmatrix}.$$

The point of doing this is that the equations in this form can be solved numerically very easily in a spreadsheet such as Excel. There are built-in functions to invert the 3×3 matrix and to multiply it into the right-hand column matrix to get the values of R_0, $a\cos\theta$, and $a\sin\theta$. Now use a very similar set of steps to write out the three equations for the separate values of the declination, d, convert to a matrix equation and solve for the values of d_0, $b\sin\theta$, and $b\cos\theta$. Notice there is a little bit of redundancy. From each set of fitted numbers I can find a value of $\sin\theta/\cos\theta$. They should, of course, be the same. If the region of space you are using is small enough, they will be very similar. If they are not, try to use a smaller grouping of stars. Once the values of R_0, d_0, θ, a and b are known they can be used to find R and d for the planet.

Using spherical coordinates

An alternative way to attack the mapping problem is to move the pixel coordinates of the stars and planet on to a sphere and then to find the rotation in three dimensions that relates the mapped pixel coordinates into right ascension and declination. The advantage of this route is that it does not contain the approximation that the celestial sphere can be treated as flat. I have been using x- and y-axes to label locations on a two-dimensional plot. The picture is easily extended to locate points in three-dimensional space. A third axis, the z-axis, is introduced perpendicular to both the x- and y-axes. These are called Cartesian coordinates, after Rene Descartes.

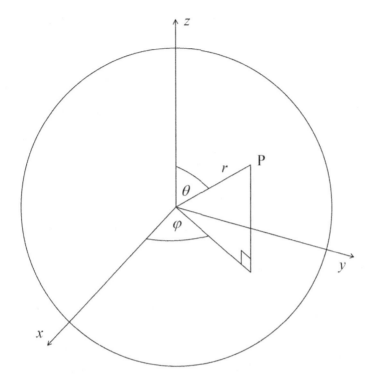

Figure 11.2 Cartesian and Spherical Coordinates

There is another system of labeling that is very useful in three-dimensions, particularly when the points can be regarded as lying on the surface of a sphere. These are called spherical coordinates. Figure 11.2 shows the relationship between the Cartesian and spherical systems of labeling. P is a point on the surface of a sphere with radius r. Its position can be described by values of the x-, y-, and z-coordinates, or, just as well, by the values of r and of the two angles θ and φ. It is helpful to be able to move back and forwards between the two coordinate systems. The coordinate system familiar to astronomers, that uses right ascension and declination to locate stars on the celestial sphere, is closely related to a spherical coordinate system except that the conventions are different. Declination is measured upwards from the equator whereas the angle θ is measured downwards from the z-axis, or polar axis, so that the declination is 90° minus the value of θ. Right ascension is traditionally measured in hours, minutes, and seconds instead of degrees. The values can be converted into degrees at a ratio of 15° for each hour of right ascension. Notice that both right ascension and the so-called azimuthal angle φ are conventionally measured in an anticlockwise

Distance measurements

sense. A point with spherical coordinates r, θ, and φ has the Cartesian coordinates

$$x = r\sin\theta\cos\varphi,$$
$$y = r\sin\theta\sin\varphi,$$
$$z = r\cos\theta.$$

The radius of the celestial sphere is not a meaningful number so I can just set it to 1. Then, for each of the three stars I can use their right ascensions and declinations to find a set of x, y, and z values. I can find a second set of x, y, and z components based on the geometry of the photograph, shown in Figure 11.3.

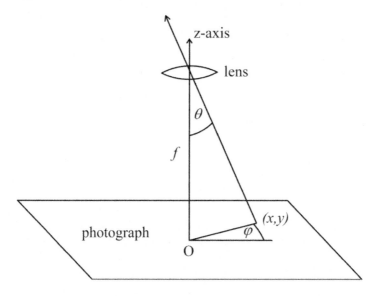

Figure 11.3 Axes based on the photograph

The object being measured is at the position (x, y). The straight-line distance of the point from the origin is given by Pythagoras's theorem as

$$s = \sqrt{x^2 + y^2}.$$

The angle φ is given by

$$\varphi = \tan^{-1}(y/x).$$

Now imagine a third axis, the z- or polar axis, using the same value of φ, perpendicular to the x- and y-axes, pointing out of the camera lens into space.

66

A line from the point on the photograph, passing through the camera lens makes an angle θ with the polar axis, with

$$\theta = \tan^{-1}(s / f).$$

I am assuming that the camera is focused on infinity so that f is the distance between the photograph and the lens. φ and θ are now the two angular coordinates of a set of spherical coordinates, and the (x, y, z) component set is completed by setting

$$z = f \tan \theta.$$

It is necessary in this approach to be less casual about the units for the components. I mentioned earlier that I scale the x and y components by a factor of the pixel size divided by the focal length of the lens, and the z component is then just $\tan\theta$. I also pointed out in the previous chapter that it is probably not good enough to use the nominal value for the focal length of the lens. You should calibrate it.

The change from the photograph-oriented set of axes to the right ascension/declination set of axes is a rotation in three-dimensional space. Manipulations of this kind are very compactly handled by means of matrices. In this case I needed a (3×3) matrix, which is an array of nine numbers in three rows and three columns. This matrix equation replaces the two trigonometric equations for R and d that I wrote out for the linear approximation method. The tactics that I used to find the matrix are very similar to my solution for the linear problem. For each row of the matrix I wrote out three simultaneous equations using the (x, y, z) for the three stars in the two coordinate systems. In each case I could solve the simultaneous equations using a matrix inversion method. Doing this for each row gave me the complete 3×3 matrix.

A matrix that describes a rotation has the form known as an orthogonal matrix. I have ignored that in finding my solution, but I can check it after the fact. An orthogonal matrix should have several identifying properties. One particular product of matrix elements, called the determinant, should have the value 1. There is a function in Excel to find the determinant so I print out its value next to the matrix on the spreadsheet. It is typically within one or two percent of the proper value.

Once I have the rotation matrix, I can feed in the pixel coordinates of any other object in the photograph and the program uses the rotation matrix to find the right ascension and declination of the object. My results

Distance measurements

are usually within 15 arc seconds for the declination and one second of time for the right ascension.

I originally set this up as a TrueBasic program, and I have used it in that form for several years. More recently I transcribed that into an Excel spreadsheet, which also works well. Both the complete and linear approximation spreadsheets can be downloaded from the web page www.GaugingtheSolarSystem.com.

Chapter 12: Distance to the Moon

I actually have tried three ways to find the diameter of the Moon and its distance from the Earth. The first is a very ancient way, used by Aristarchus of Samos, around 280 BC. He used the idea that the dark shadow seen on the Moon during a lunar eclipse is the shadow of the Earth. Other astronomers had earlier pointed out the curved shape of the shadow as a proof that the Earth was a sphere. Aristarchus used this as a quantitative method of finding the relative sizes of the Moon and the Earth. His procedure is described in "A History of Astronomy" by Pannekoek. His first step was to estimate the diameter of the shadow compared with the diameter of the Moon. According to Pannekoek, he came up with a ratio of two to one. This is not a particularly accurate answer, but Aristarchus got lucky in his analysis. He combined this result with his earlier finding that the distance from the Earth to the Sun is nineteen times the distance from the Earth to the Moon. This is, of course, a serious underestimate, but in the calculation of the size of the Moon the two errors compensated and his final value for the size of the Moon was quite respectable.

I wanted to reproduce this method, modified to use the idea that the Sun is very distant from the Earth. I wrote a short article on this for a physics educational journal many years ago. (American Journal of Physics, volume 57, page 351, April 1989). At that time I relied on published photographs of the Moon during the partial phases of an eclipse, but I had always intended to do the whole job myself. During the lunar eclipse of December 21, 2010, I took a number of digital photographs of the partial phases of the eclipse, using the zoom lens set at 300 mm focal length. The radius of the image of the Moon on my photographs was 223 pixels. Next, in the computer, I made a template containing a series of concentric circles with radii in steps of fifty pixels, with a transparent background. I superimposed the template over each of the photographs, as a separate layer, and slid this layer over the photograph until the edge of the shadow best fitted one of the circles on the template. There is some uncertainty here because the edge of the shadow is fuzzy. A reason I had taken several photographs is that I could require that the same circle fitted all of them. I decided that 500 pixels was definitely too small, 700 pixels was too large, 600 was just about right, but 550 or 650 pixels might do. So my result is 600 ± 50 pixels. The ratio of the size of the shadow to the size of the Moon's disk is therefore $(600 \pm 50) / 223 = 2.7 \pm 0.2$. Call this number R. In appendix 1 at the end of the chapter, I derive the formula that the diameter of the Moon divided by the diameter of the Earth is

$$\frac{D_M}{D_E} = \frac{1}{R+1}.$$

Distance measurements

The number 1 in the denominator is not exact, but is close enough. Putting in my measured value for R

$$\frac{D_M}{D_E} = \frac{1}{3.7 \pm 0.2} = 0.27 \pm 0.02.$$

I have been a little cautious with the uncertainty. It actually calculates out as 0.015. My answer is very close to the handbook value.

Once I have the diameter of the Moon, I can find the distance of the Moon from the Earth, still in units of the Earth's diameter, by using the angular size of the Moon. I describe my angular size measurements in chapter 19, although it is very well known that the Moon is about half a degree across. For such a small angle, the angular size in radians is just the diameter of the Moon divided by the distance from the observer. There is a small complication. The Moon is sufficiently close to us, compared with the radius of the Earth, that its angular size seen from the surface of the Earth is larger than an observer at the position of the center of the Earth would see. Indeed, the angular size of the Moon seen by an observer on the Earth's surface does change during the course of an evening, as the Earth turns. I shall use this as a basis for my next method to determine the distance. When I took the measurements described in chapter 19, I was already sensitized to this behavior, and I tried to take the full moon measurements when the Moon was high in the sky. The average of my measurements of the full Moon is 31'27" = 0.00915 radians. The average distance from the Moon to the surface of the Earth is then

$$\frac{0.27 \pm 0.015}{0.00915} = 29.5 \pm 1.6.$$

This is the distance in units of the Earth's diameter. In terms of the radius it is just twice this, or 59 ± 3, and the distance from the Moon to the Earth's center is about one Earth radius greater than this, or 60 ± 3. Alternatively, my measurements of the size of the new Moon should correspond closely to the angular size seen by an observer at the Earth's center, since the distance to the Moon is very similar in the two cases. Unfortunately, I don't have enough of those to give an accurate average. The average of my measurements of the quarter and three-quarter moons is 31' 8" = 0.00906 radians. If I use this value as a substitute, I calculate the distance of the Moon from the Earth's center directly as 30 Earth diameters, or 60 Earth radii, once again.

As in several other places in this book, if you manage to push the precision of the measurements higher, there are small corrections that need to be taken into account. In this case, there is substantial evidence that the refraction of sunlight in the Earth's atmosphere has the effect of increasing slightly the diameter of the Earth's shadow on the Moon during a lunar eclipse. A discussion of the effect was given in an article in Sky and Telescope magazine, the issue of September 1996, pp 98-100, by Erich Karkoschka. The correction is only one or two percent, and I have ignored it since the relative uncertainty of my observation is much larger, but if you do better than me you might need to include it.

In this method, it was the diameter of the Moon that was determined directly and its distance from the Earth then follows, almost trivially, from the angular size of the Moon. If I use the average angular size I get the average distance. The other two methods give the distance to the Moon directly. A drawback to this is that the distance to the Moon is continually changing and you have to accept whatever it was at that time. I backed into the second method from a different set of ideas. There is a well-known effect, I shall call it the full-moon effect, that the full moon sometimes looks enormous when you see it close to the horizon. Many

Figure 12.1 The Moon close to the horizon and high in the sky

people assume that there is some optical reason for this involving the curvature of the Earth's atmosphere, but the standard explanation is that the effect is purely psychological. The brain is responsible for the apparent large size of the Moon, possibly because the closeness of the Moon to the horizon allows you to compare its size with familiar objects such as trees and buildings. I quite often talk to groups of non-astronomers, either young people or adults, and I thought it would be interesting to make a pair of photographs, taken on the same evening, showing the Moon when it was close to the horizon and when it was high in the sky. I will claim credit for realizing what was going to happen before I had finished the project. Far from the Moon looking larger when it was near the horizon, the image taken when the Moon was high in the sky was actually slightly larger!

In Figure 12.1, the left hand image is of the Moon close to the horizon (with power lines in front of it) and the right hand image was taken when the Moon was close to the meridian. The two photographs were taken just a few hours apart. A qualitative reason for the effect is shown in Figure 12.2. The circle is the Earth seen from above the North Pole. When the Moon is on the horizon, the line of sight from the observer to the Moon is tangential to the Earth, and the distance from the observer to the Moon is quite close to the distance from the center of the Earth to the Moon. When the Moon is directly overhead, the Moon is actually closer to the observer by the radius of the Earth. Of course, these pictures are two-dimensional. The diagram applies to an observer at the equator, when the Moon rises exactly in the East. For an observer who lives outside of the tropics, the Moon is never directly overhead even in the middle of the night. The effect is reduced but it is still there. So I realized that I could get a value for the distance to the Moon by measuring the difference in size of the images on the two photographs. There are two difficulties in the experiment. The first is that the distance of the Moon from the center of the Earth does change slightly during the course of the evening. There are plenty of computer programs that will give you the numbers, but they also know all about the value I am trying to measure, so that seems cheating. My solution was to take a series of photographs, at about the same time in the evening, for a few days before and after the evening of full moon. From them, I could estimate the part of the change in size that came from this motion. The second problem was the viewing conditions at my home. I found that the photographs taken at moonrise showed a lot of distortion because I was looking through dirty air. I would typically take several photographs and there was a spread in values of the diameter of the Moon images. Of course, there is a real distortion of the shape of the images caused by the atmosphere. The images are distinctly non-circular, squashed vertically. I assume that the horizontal diameter is the true image size. Anyway, I repeated the experiment on several full-moon evenings, using my usual set up at the prime focus of the 90 mm refractor. Finally, on February 10, 2012, I produced a pair of good quality photographs. The increase in size of the image was 30 ± 3 arc seconds, and I could estimate

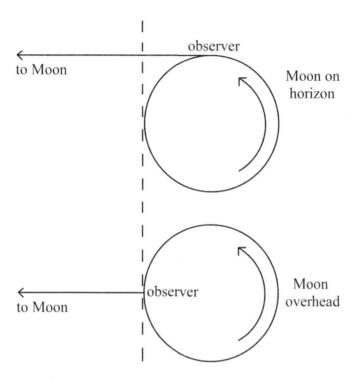

Figure 12.2 Change of distance from observer to Moon as Earth rotates

that 4 seconds of that was caused by the motion of the Moon towards the Earth. Even on that evening I had to compromise a little. The weather forecast for the night predicted clouds, so I took the late night picture almost an hour before the Moon actually transited.

The equations for calculating the two distances for the realistic conditions: the Moon not quite on the horizon in the first photograph and not quite overhead for the second, are given in Appendix 2 at the end of the chapter. They are quite complicated so I resorted to a numerical solution. In a short computer program I stepped through a range of values of the ratio of the distance to the Moon divided by the radius of the Earth, and for each value I calculated the two angular sizes and found their difference and their ratio. I could then pick out the value of R/R_E that gave a difference of 26 seconds and also find the uncertainty in the ratio corresponding to a 3 second uncertainty in the difference. The number that came out was 59 ± 7. This is the distance from the center of the Earth to the Moon in units of the radius of the Earth. The value is close to the value I got from the first method but with a larger uncertainty. Actually, from a handbook I can find that the Moon was moving towards perigee, so this answer is a bit higher than it should have been on that date, but the large uncertainty covers that.

The third method I have tried is to leverage my value for the radius of the Earth to find the distance to the Moon directly. So far it hasn't given me a result comparable with the other two methods, but I think it is neat that should exist at all. I based it initially on a project described in "A Workbook for Astronomy" by Jerry Waxman. During the course of an evening, the Moon moves along its orbit at an average angular speed of just over 13° per day, or 0.55° per hour, moving from west to east. However, at the same time, the Earth rotates on its axis and an observer on the Earth's surface is also moved from west to east, at 15° per hour. If the Moon was not moving relative to the stars, this movement of the observer would cause the Moon to appear to move backwards, from east to west, by a small amount. The combined effect is that the angular speed of the Moon as viewed by that observer is reduced. If the Moon is on your meridian, its angular speed is reduced by an amount

$$15^{o} \text{ / hr } \times \frac{\text{distance of observer from Earth's rotation axis}}{\text{distance of observer from Moon}},$$

which works out to about 0.19°/hr at my latitude of 40°N. This is an example of a parallax effect, similar to the retrograde motions of the superior planets. The experiment is to pick an evening when the Moon is reasonably close to a star that is close to the ecliptic, and to measure the separation of the Moon and the star through the evening. The rate of change of their separation, in degrees per hour, is less than the actual angular speed of the Moon relative to the stars and the difference can be used to calculate the distance of the Moon. There is (as usual, I hear you say) one complication that has to be taken into account. The Moon moves in an approximately elliptical orbit round the Earth, and its angular speed varies from point to point in the orbit. The 13° is an average value. At perigee, the number is larger by about 11% and at apogee it is smaller in the same ratio. The first time I tried this measurement, I hadn't thought about that. While the answer I got was not bad, I later went back and re-analyzed the numbers and improved the result, but it still lags behind the other two approaches. I shall keep trying.

Appendix 1: The relative size of the Moon

During a lunar eclipse, the Moon moves through the cone of shadow of the Earth. From my measurements during an eclipse, I found that, at the distance of the Moon, the conical shadow had a diameter 2.7 times the diameter of the Moon. If the shadow was not tapered, this would imply that the diameter of the Earth was 2.7 times that of the Moon. Because the shadow does taper, the ratio is greater than this. A very not-to-scale picture looks like this

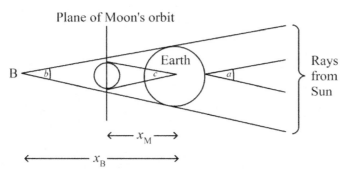

Figure 12.3 Geometry of a Lunar Eclipse

In fact, the rays coming in on the right from the Sun are almost parallel, and furthermore, the Sun is so far away on the right of the picture that the extra distance that the point B is from the Sun does not change the slope of the rays. That is, I can assume that the angles a and b are equal. c is the angle subtended by the Moon. There is no obvious reason why that should have the same value, but, at the present time, it is very closely equal to the other angles, leading to the beautiful effects at a total eclipse of the Sun. The three angles are all approximately 0.5°, although the actual value doesn't matter. The angle b is related to the diameter of the Earth, D_E and the distance x_B of point B from the center of the Earth by

$$b = \frac{D_E}{x_B} \text{ or } x_B = \frac{D_E}{b}.$$

This is valid because b is such a small angle: if it was larger I would need to use a trig function. In the same way, the distance x_M of the Moon from the Earth is

$$c = \frac{D_M}{x_M} \text{ or } x_M = \frac{D_M}{c}.$$

Distance measurements

The diameter of the earth's shadow at the distance of the Moon is

$$b(x_B - x_M) = D_E - \left(\frac{b}{c}\right)D_M \approx D_E - D_M.$$

So, finally, the size of the Earth's shadow is larger than the size of the Moon by a factor

$$R = \frac{D_E - D_M}{D_M}.$$

A small rearrangement gives the equation I need

$$\frac{D_M}{D_E} = \frac{1}{R+1}.$$

The "1" in the denominator did depend on the approximation that the angles b and c were equal, but this is a really good approximation.

Appendix 2: The full moon effect

For a general position on the Earth's surface, i.e. not on the equator, and for a general position of the Moon in the sky, the distance between the Moon and an observer on the Earth's surface depends on the right ascension, RA, and declination, δ of the Moon, and the time of the Moon from transit at the observer's location. This last number is called the local hour angle, or LHA. Duffett-Smith has the necessary subroutines for calculating it from the right ascension of the Moon and the observer's longitude, either as TrueBasic subroutines or as Excel pages. The complete equations are a bit messy, so jump to the bottom line if you feel like it.

Choose a set of axes centered at the center of the Earth so that the z-axis runs along the Earth's rotation axis, through the North Pole, and the x-axis emerges from the equator at the observer's longitude. Then the x, y, and z coordinates of an observer on the Earth's surface at a distance R_E from the origin and at latitude λ are

$$x_{obs} = R_E \cos(\lambda),$$
$$y_{obs} = 0,$$
$$z_{obs} = R_E \sin(\lambda).$$

If the Moon is at distance R, its coordinates are

$$x_M = R\cos(\text{LHA})\cos(\delta),$$
$$y_M = R\sin(\text{LHA})\cos(\delta),$$
$$z_M = R\sin(\delta).$$

The distance between the observer and the Moon is

$$d = \sqrt{\left(x_M - x_{obs}\right)^2 + \left(y_M - y_{obs}\right)^2 + \left(z_M - z_{obs}\right)^2}.$$

This is a three-dimensional form of Pythagoras's theorem. When the expressions for the various components are put in, and the equation is cleaned up a bit using trigonometric identities, the result is

$$d = R\sqrt{1 - \frac{2R_E}{R}\cos(\text{LHA})\cos(\delta)\cos(\lambda) - \frac{2R_E}{R}\sin(\delta)\sin(\lambda) + \frac{R_E^2}{R^2}}.$$

Distance measurements

Let's use the letter S to represent that slightly messy square root. For any value of the ratio R/R_E, I can calculate this for the two times when I took photographs and get two values for the square root, S_1 and S_2. I do make some approximations to simplify life. First, the right ascension and declination of the Moon do change a bit between the two photographs, but I ignore that. More importantly, the distance of the Moon from the Earth also changes. I had estimated this change as producing a 4 second change in the size of the image. I assume that I can handle that sufficiently well by subtracting 4 seconds from the size of the second image but then using the same value of R in the equation at both times.

We're almost there. To be specific about my results, I measured the angular size of the Moon as 1914 seconds on the photograph taken just after moonrise, and as 1944 seconds on the photograph taken close to transit. The difference is 30 seconds, and each value had an uncertainty of 1 or 2 seconds, so I am going to say the difference is (30 ± 3) seconds. Of this, 4 seconds comes from the motion of the Moon towards the Earth, so I subtract that to get 1940 seconds for the late night size and (26 ± 3) seconds for the difference. The two angular sizes are D_M/RS_1 and D_M/RS_2, where D_M is the diameter of the Moon. If I take the ratio of the two sizes, D_M and R both cancel and what is left is

$$\frac{S_1}{S_2} = \frac{1940}{1914} = 1.0136 \pm 0.0016.$$

I spent some time looking for analytic solutions to this, and then thought "Why bother?" It took me ten minutes to write a TrueBasic program that calculated this ratio for a series of closely spaced values of R/R_E and I could pick out the correct value and the error limits. You could do it just as easily in a spreadsheet.

Chapter 13: Orbits of the Inferior Planets

The planets go round the Sun in elliptical orbits, but the eccentricities of the ellipses are all small. That is, the ellipses are almost circles. A vivid illustration of this is that a circle drawn on an 8 ½ x 11 inch sheet of paper with a pencil of normal sharpness will contain within the thickness of the pencil line an ellipse with the eccentricity of the Earth's orbit. (I like this example but there is a catch to it. In the diagram the Sun will visibly not be at the center of the circle. It will be a couple of millimeters from the center, at one of the foci of the ellipse.) Certainly, for a first try at finding the sizes of the orbits, it is reasonable to approximate them by circles. For the two inferior planets, the points of maximum elongation in their orbits offer a simple geometry for determining their relative sizes. At maximum elongation, the line from the Earth to the planet is a tangent to the circular orbit of the planet, and the planet is exactly half illuminated.

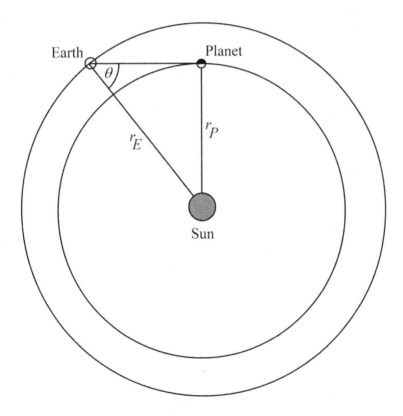

Figure 13.1 Maximum eastern elongation of an inferior planet

79

Distance measurements

The angle between this tangent line and the line from the Earth to the Sun is given by

$$\sin(\theta) = \frac{r_P}{r_E}.$$

The distance of the planet from the Sun is then

$$r_P = r_E \sin(\theta).$$

The average distance r_E of the Earth from the Sun is an astronomical unit (a.u.), so that, in these units r_E is defined to be 1. Then, for the two inferior planets

$$r_P = \sin(\theta) \text{ in a.u.}$$

However, if the uncertainties of the measurements are better than a few percent, it is worth finding more accurately the distance of the Earth from the Sun at the time of the observation. You can get this from one of many computer programs, but, as I describe later, I eventually got tired of relying on other people's values and wrote a program of my own to solve Kepler's equation. It doesn't have the accuracy of a program such as the U.S. Naval Observatory Multiyear Interactive Computer Almanac (MICA), but it is good enough for this purpose, and I feel better using it!

Measuring the angle θ is not completely trivial. First, the measurement has to be made at the time of greatest elongation. This requires either that you measure the angle separating the planet from the Sun over a period of time and find the largest value, or that you swallow your pride and look up the date of maximum elongation in an almanac. I give both alternatives because I have used the first for Venus but the second for Mercury. Mercury is quite elusive and it also moves through maximum elongation over a period of only a few days, so you have to be lucky with the weather. Venus spends a much longer time in either the eastern or western sky and there are better opportunities for tracking it.

The second point is that the Sun is much brighter than either of the planets. I have not yet been able to see Mercury while the Sun was in the sky although I have seen Venus before sunset, after sunrise, and, indeed, in the middle of the day. Of course, looking at the Sun with the naked eye is a dangerous thing to do, so you need to devise a safe procedure. I can use my small sextant, which does have safe filters for viewing the Sun, to measure the angle between Venus and the Sun while both are in the sky, but I got a more complete set of measurements using the photographic method after sunset or before sunrise.

I have tracked four eastern and three western elongations of Venus. I take a series of photographs and use the program described earlier to find the right ascension and declination of Venus. I can then calculate its elongation from the Sun using some of Duffett-Smith's subroutines. The results are generally on a smooth curve with deviations of a few hundredths of a degree. The one limitation is that the photographs have to include at least three identifiable stars, to be used as input for my analysis program, so I have to wait for the sky to darken enough for them to show, and Venus was sometimes getting low in the sky by then. A typical result is shown in Figure 13.2.

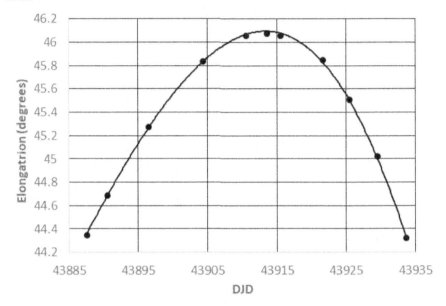

Figure 13.2 Eastern elongation of Venus, 2020

To locate the maximum as precisely as possible I use Excel to fit a polynomial curve to the maximum. This allows me to calculate the position of the maximum and also corrects for the small experimental deviations from a smooth curve. I found that the maximum elongation of Venus from the Sun was 46.1°. The formula including the actual distance of the Earth from the Sun then gives the radius of Venus's orbit as 0.718 a.u. The table shows the values for all seven elongations.

The distances of Venus from the Sun are in astronomical units, While the differences are small, I believe the values are accurate to at least three decimal places and simply indicate that the orbit of Venus is almost a circle. The best I can do at the moment is to take the average of them as my

Distance measurements

best measurement of the orbit size. In the table in the introduction I wrote it as 0.723 ± 0.004 a.u.

Date	Venus Maximum Elongation	Distance from Sun
8/19/2010	45.976°	0.7277
1/7/2011	46.972°	0.7188
6/6/2015	45.394°	0.7225
10/26/2015	46.431°	0.7203
1/12/2017	47.152°	0.7211
6/3/2017	45.855°	0.7279
3/24/2020	46.081°	0.7184

For Mercury, I have the four photographic records of elongations that I used in Chapter 6. I also mentioned there some less precise, earlier observations that I had made. I have resurrected them to include in the list of distances for Mercury,

Date	Mercury Maximum Elongation	Distance from Sun
10/6/2009	18.5°	0.32
1/27/2010	27°	0.45
4/10/2010	19.4°	0.33
9/20/2010	17.9°	0.308
5/7/2015	21.22°	0.365
4/18/2016	19.95°	0.343
9/28/2016	17.685°	0.307

The calculated distances range from 0.31 a.u. to 0.45 a.u. I do not believe that this variation is due to experimental uncertainty but that the range is caused by the eccentricity of the Mercury's orbit. A wildly optimistic interpretation is that if the average distance is r_M and the eccentricity is e, then the highest and lowest values I have measured are getting close to $r_M \times (1 - e)$ and $r_M \times (1 + e)$. Solving on this basis gives a value of 0.38 a.u. for the average distance and 0.18 for the eccentricity. These are amazingly close to the handbook values considering the crudeness of the data.

As the Earth and the other planets revolve around the Sun there are times when the Earth passes between the Sun and a superior planet, or an inferior planet moves between the Sun and the Earth. In the first case the outer planet is said to be in opposition and in the second case the inner planet is in inferior conjunction. In either case, the planet as viewed from the Earth appears briefly to move backwards relative to the stars, westward instead of its usual eastward motion. In the next chapter I shall use the so-called retrograde motion of the outer planets to determine their orbital parameters. I have followed all of them except for Neptune several times. The retrograde motion of the inner planets is much more difficult to observe because they are passing close to the Sun. Early in 2017 I was tracking the Eastern elongation of Venus and realized that its orbit was about as high above the ecliptic as it ever gets, and I decided to try for the retrograde motion.

Inferior conjunction was in fact on March 25th. The last day when I was able to take a usable evening photograph was March 16th, and the first morning when I could detect stars on a photograph was April 2nd. I was able to see Venus several times between these dates, but in the photographs the sky was too bright to show stars. A factor that contributed to the difficulty was that Venus performed its retrograde motion in a region of sky with no bright stars in it. The stars that I was able to use were mostly of the 4th and 5th magnitudes. The results for the right ascension and declination, in the photographs taken close to inferior conjunction, were not quite as accurate as the values I have been used to getting for the superior planets. This could be in part because the stars that I could identify were all much higher in the sky than Venus, so I was making a long extrapolation in the coordinate transformation. Also, I knew that Venus appeared as a narrow crescent when seen through a telescope, whereas the image in my camera, while not exactly circular, showed no trace of the crescent. Presumably the numbers that came out of my spreadsheet corresponded to some sort of weighted center of the illuminated part. Fortunately, the errors are in fact quite small on the scale of the graphs I have plotted.

A graph of the declination versus the right ascension shows the shape of the loop much as it would appear to an observer. I have plotted the right ascension in degrees, rather than the usual hours, so that the horizontal and vertical scales are comparable. Remember that right ascension conventionally increases from west to east, or right to left in the northern hemisphere, so the loop would be flipped. The two reversals of the direction of motion are obvious. Compare this graph with Figure 14.1 for Mars in the next chapter. The reversals occurred around March 2nd and April 13th, but that is not clear in this type of plot. The graph of ecliptic longitude as a function of time, shown in Figure 13.4, is a reliable indicator of the time scale of the loop. I have actually plotted the time as days after January 1st, 1900, which is a version of the Julian day numbers, and I have added 360° to some of the longitudes to give a continuous curve.

Distance measurements

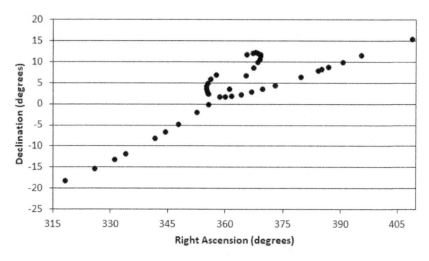

Figure 13.3 Retrograde loop of Venus, 2017

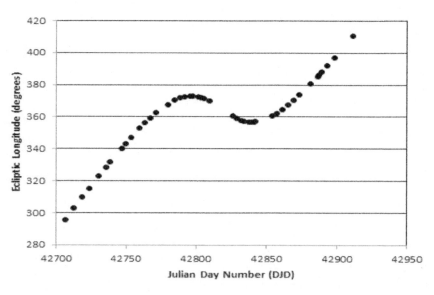

Figure 13.4 Ecliptic longitude of Venus. 2017

The points where the two curves change direction are called "stationary points" because Venus is apparently stationary with respect to the stars. These are not the same as the maximum eastern and western elongations. The maximum elongations are points when Venus is apparently stationary with respect to the Sun. At the "true" stationary points Venus is falling behind the Sun, as it is viewed from the Earth. (There is a report of this at www.skyandtelescope.com/observing/stargazers-corner/tracking-venus-retrograde-motion/)

84

Chapter 14: Orbits of the Superior Planets

To find the orbital parameters of the planets whose orbits are outside the Earth's, I made a careful series of observations of their apparent positions relative to the stars close to the time of opposition. The parallax effect arising from the motion of the Earth causes the planets apparently to reverse their motion. This is the retrograde motion effect that caused such headaches for the early astronomers and led to the development of the epicycle models for their orbits. In the previous chapter I described my measurement of the retrograde loop of Venus.

I described in chapter 11 the different methods I have used to find the right ascensions and declinations of objects in the sky, leading to a photographic method that has been very satisfactory. I developed these methods in the first place to apply to this particular set of observations. I followed the retrograde loop of Mars in 2007 – 2008 using some of the simpler techniques and then followed its motion in 2009 – 2010 mainly using the photographic method. The plot of declination versus right ascension shows the shape of the loop.

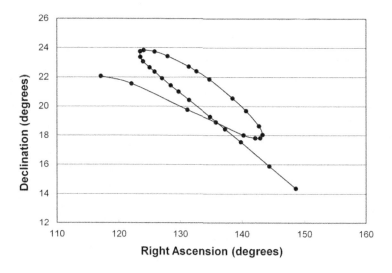

Figure 14.1 Retrograde loop of Mars 2009 - 2010

As I have said earlier, I tend to restrict my mental picture of the motion of the planets as taking place in a plane, and the ecliptic longitude seems the best number to plot as a function of time. This is shown in Figure 14.2.

Distance measurements

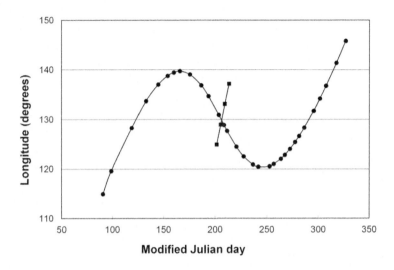

Figure 14.2 Ecliptic longitude of Mars during retrograde loop 2009 - 2010

The short line with square symbols is the longitude of the Sun minus 180°. The intersection of the two lines identifies the moment of opposition. I find that I can usually locate it to well within an hour. On the diagram for Saturn, later, I have also plotted a short line for the longitude of the Sun plus 90°. The crossing of this line with the line for the planet identifies the time when the planet is at quadrature, when the line joining the planet to the Earth makes an angle of 90° with the line joining the Earth to the Sun. I initially took the numbers for the Sun from the Observer's Handbook. As I mention later, I did develop for myself a short computer program that solves Kepler's equation for the Earth to give the distance from the Earth to the Sun and the longitude of the Sun through the year. My program isn't quite as accurate as the Observer's Handbook or MICA, but it is close enough that I use it for most of the fitting described below, and also feel comfortable using the handbook values if the higher accuracy seems necessary.

For the other planets I did take a short cut. Observation of the complete loop does take quite a long time and, looking at the curve of ecliptic longitude for Mars, I decided that there would be sufficient information if I started the observations just before the time of opposition and traced out only the second half of the loop (which does occur in the evenings, rather than the early hours of the morning). I also mentioned earlier that by happy chance Uranus and Jupiter were close together in the sky for the oppositions of 2010, so that one set of photographs included both planets. The graph of their ecliptic longitudes is shown in Figure 14.3.

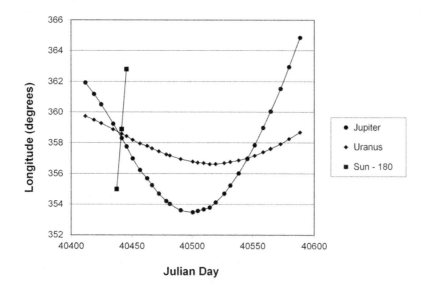

Figure 14.3 Ecliptic longitudes of Jupiter and Uranus 2010

Finally, Figure 14.4 shows my results for Saturn, taken in 2011.

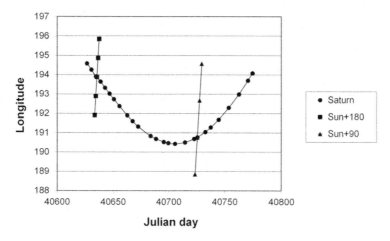

Figure 14.4 Ecliptic longitude of Saturn 2011

I measured retrograde loops for Mars in 2007 – 2008 and 2009 – 2010, Jupiter in 2009, 2010 and 2011, Saturn in 2010 and 2011, and Uranus in 2010 and 2011. Then I took a few years break but returned to the job to measure complete curves for Mars in 2012, 2014, and 2016, and for Jupiter in the Fall of 2014 and Spring of 2015, and in 2018, plus a half curve in 2016; also second-half curves for Saturn in 2015, 2016 and 2020, with a complete curve in 2018, and half-curves for Uranus in 2016 and 2018. A small comment is that the planets spend more time in retrograde loops than I had appreciated. Until I made these measurements I had always pictured the retrograde motion as a small glitch in a predominantly forward motion. But the actual retrograde motion occupies about 81 days for Mars, 118 days for Jupiter, 140 days for Saturn and 152 days for Uranus. The part of the curve I have used for my analysis includes the parts of the forward motion that cover the same ground as the retrograde motion, the part between the two points labeled A in the diagram in the appendix. The total time including these parts is 168 days for Mars, 228 days for Jupiter, 272 days for Saturn, and 291 days for Uranus. The chances are that if you look at one of the superior planets during the evening it is probably going through this loop.

There is clearly a lot of information in these diagrams, and the question is how best to extract it. The extraction of the planetary parameters is probably more complicated than the actual recording of the data. This is one place where you should really stop and think your own way through the analysis. You might also look at the rather different methods used by Dr. Clark, in his book listed in the bibliography. I found myself going deeper and deeper into the analysis. You might not want to do that, although remember my comment at the beginning of this book that we get a lot of cloudy nights when you need something to do other than looking at stars, so let me start with the easiest approach.

In the appendix, I produce an equation to describe the line-of-sight angle, θ, of the planet as a function of time around opposition. It involves the angle turned through by the Earth since the time of opposition, α_E, and the angle turned through by the planet, α_P, and the distances of the planet and the Earth from the Sun, r_P and r_E. In terms of the angular velocities of the Earth and planet, the two angles are equal to $\omega_E t$ and $\omega_P t$. The easiest approach is to concentrate on a small number of special points on the curves for which the math expressions are relatively simple. One or more parameters can then be fitted to them. A more ambitious method of attack is to perform a non-linear least squares fit of the whole curve to all of these numbers, and I shall describe that later.

If I assume that planets move at constant angular speed, the angular velocities are related to the sidereal periods, T, so that

$$\alpha_E = 360° \times \frac{t}{T_E} \text{ and } \alpha_P = 360° \times \frac{t}{T_P}.$$

If the orbits are circular, r_E is just one astronomical unit. I have found the average periods of the planets more-or-less accurately, as described in chapter 7, mainly using the times of opposition from these results. If I suppose that I can use these values, there is only one number in the equation for $\tan(\theta)$ that I don't know the value of, and that is r_P. I can now use any one of the special values that I work out in the appendix to find an estimate of the distance of the planet from the Sun.

To illustrate the pros and cons of the method, let me apply it to my results for Jupiter in 2010. The easiest special points to work with are the ones I labeled A in the figure in the appendix. They are the points on either side of opposition when the planet is in the same position as at opposition. If t_A is the time of one of these points before or after opposition, the radius of the planet's orbit is given by

$$r_P = r_E \frac{\sin(360° \times t_A / T_E)}{\sin(360° \times t_A / T_P)}.$$

If I suppose that the distance of the Earth from the Sun is 1 astronomical unit and that the period of the Earth is 1 year, this simplifies (with all the times measured in years) to

$$r_P = \frac{\sin(360° \times t_A)}{\sin(360° \times t_A / T_P)} \text{ in a.u.}$$

For example, from the curve that I measured for Jupiter in 2010, the time t_A can be obtained as 113.81 days, or 0.3116 years, and the value for the sidereal period that I found in chapter 7 was 11.91 years. (Not all these figures are meaningful, but I usually carry more than I think I shall need and round the answer at the end.) With these values, the radius of Jupiter's orbit comes out to be 5.64 astronomical units.

Other special values are the time from opposition to a stationary point, S, and the angular speed of the planet at the time of opposition. The formulas for these are a little more fussy, and are in the appendix. Another convenient point, not directly connected with the retrograde motion, is the quadrature position of the planet. This is similar to the position of maximum elongation of the inferior planets, and is indeed the point at which the Earth observed from the outer planet would be at maximum elongation from the Sun. You can locate it on the graphs of the ecliptic longitude of the planets by plotting the longitude of the Sun plus or minus 90°, as I did for Saturn. If you assume that the planet, or even more importantly, the Earth, moves at a

Distance measurements

uniform speed, you can turn this time of quadrature into an angle between the two lines connecting the Earth and the planet to the Sun, and the picture is then the same as I used for the inferior planets, except that the roles of the planet and the Earth are reversed. In the case of Jupiter in 2010, the time for the planet to reach eastern quadrature was 86.388 days, or 0.2365 years, after opposition. In that time, assuming uniform motion, the Earth has moved through 85.146° and Jupiter has moved through 7.095°. The formula given in the appendix then gives the distance of Jupiter from the Sun as

$$\frac{1}{\cos(85.146 - 7.095)} = 4.83 \text{ a.u.}$$

This differs by about 20% from the value I got using the A point. If you try the other two special cases that are described in the appendix, you will get two more values. The problem is that if you try all four, you will get four slightly different answers, partly because experimental results always have some uncertainty but mainly because the assumptions I have made are over-simplifications; the planets do not move in circular orbits at constant speeds in the plane of the ecliptic.

The values that I was using give a clue as to how the calculation can be improved. The time from opposition to the point A was about 114 days. This is almost four months or a third of a year. This is only a small fraction of an orbit for Jupiter so it is reasonable to assume that Jupiter is moving at a constant speed. However, it is over-optimistic to treat the motion of the Earth as taking place at constant speed, or at a fixed distance from the Sun. From an almanac or one of Duffett-Smith's computer programs I found that the angle actually turned through by the Earth in this period was actually 114.35° and that the distance of the Earth from the Sun at that time was 0.9835 a.u. Similarly, at quadrature the Earth has turned through 86.406°, instead of 85.146° for uniform motion, and it is 0.9854 a.u. from the Sun. If I put these numbers into the formulas, I calculate the two values for the distance of Jupiter from the Sun as 5.556 a.u. and 5.305 a.u. respectively. The difference is now just under 5%. Finally, you can include the possibility that Jupiter's angular speed is different from the average value. I incorporate this by changing the value I am using for the period of Jupiter, but that is just a convenient way of parameterizing its instantaneous speed. An easy way to do this is to make a small spreadsheet that calculates the distances from the two formulas, to run it for a closely spaced series of values of the period, and to find the value of the period for which the two distances are equal. That way, I found that when the period was 10.63 years both formulas gave the distance as 4.89 a.u. (Notice that this had quite a large effect. The new answer is outside the range of the two previous values.) My own TrueBasic version included all four of the special points and this shifted the results slightly to 4.95 ± 0.04 a.u. for the distance and 10.8 ± 0.2 for the instantaneous period. I think this is the best you can do with the special points method. For comparison, the procedure of fitting the parameters to the whole curve gave a value of 4.95 ± 0.02 a.u. for the distance of Jupiter from the Sun, and an instantaneous period of 10.77 ± 0.08 years. The answers are very similar but I prefer the overall fitting method because it is making use of the whole curve rather than a few selected points. For my own analysis, I used both approaches in parallel, the overall least-squares fit to the curves, and choosing special points, but in each case I gradually relaxed the assumptions.

In the least squares approach I again began by adjusting only the value of the distance of the planet from the Sun. The simplest improvements to make were to vary the period of the planet from its average value and to use values for the Earth of its speed and distance from the Sun that were appropriate to the opposition point. These helped a lot, but I felt I could do better. My next step was to treat the position of the Earth quite a bit more carefully, both its distance from the Sun and its angular velocity. Initially, I took the values from an almanac or from a commercial computer program, but I was not happy with myself for doing that. Finally, I wrote a subroutine to calculate both of these as functions of time from perihelion. The equation describing the elliptical orbit of the Earth is called Kepler's equation. It is given, for example, in Meeus's book "Astronomical Algorithms" and in Duffett Smith's book "Astronomical Calculations with your Personal Computer". I was able to base my version mainly on numbers I had measured, described in chapter 19. It is not quite as accurate as Duffett-Smith's program but I felt more comfortable using it! This gave a visible improvement.

I have described the general idea of the non-linear least squares method in appendix D at the end of the book. Using it, I can vary all or some of the parameters to give a best overall fit. One problem in performing a non-linear least-squares fit is that the process has to be iterated. The values calculated for each parameter depend on the values you start with. In a further iteration starting from the most recent values, there are still further shifts in the parameters. This happens because the parameters are correlated; changing the value of one of them alters the others. I convinced myself that the ratio r_P^2 / T_P would be less correlated to T_P than would r_P itself, so that is what I used as an independent variable in the final version of the fitting program. The two methods gave results that agreed within the uncertainties, but I preferred the least-squares fitting method because the uncertainties were less subjective and because it makes full use of all the measured points and not just the special values.

The best fitted curve for the opposition of Jupiter in 2014 – 2015 looks like this:

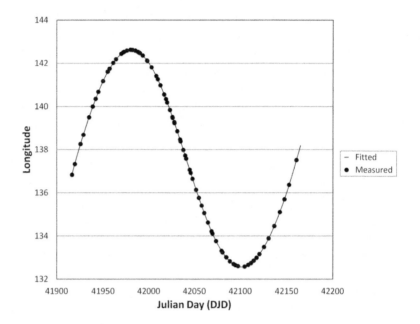

Figure 14.5 Fitted retrograde curve for Jupiter 2015

In the other plots in this chapter, the line was just a smooth line through the points, drawn by Excel. In this case, the line is separately calculated, based on the best fitted parameters. On the scale of this plot, the fit is pretty much perfect! I have since tracked Jupiter through two more oppositions in 2016 and 2018.

The values for the distance of the planet from the Sun and the instantaneous values of the period are shown in the tables below for the various planets. At the end of chapter 7, on finding the average sidereal periods of the planets, I mentioned that I had made use of a set of approximate values found by a different method. They were the values of T_P in this table. While they are not the average values for a complete revolution, they are surely quite close. The final step is to combine these two sets of numbers, the instantaneous values at opposition and the values for the average periods that I had found earlier. In the process I get to check Kepler's second law, which states that the planets move around their elliptical orbits, changing their speeds as they move so that they sweep out equal angles in equal times. In more modern terms, this is the same as saying that the quantity ωr^2 stays constant, where ω is the angular velocity, or that the ratio

R_P^2/T_P stays constant, where T_P is what I have called the instantaneous period. R_P is the fitted value of r_P at opposition. Notice that the six sets of results for Jupiter all give the same value of R_P^2/T_P within the uncertainties and the sets of values for each of the other planets also agree. This is my verification of Kepler's second law.

Results for Mars:

Date	Ecliptic longitude at opposition	Distance from Sun at opposition, R_P, (a.u.)	Instantaneous period, T_P (years)	R_P^2/T_P
12/24/2007	92.4702°	1.607 ± 0.011	2.079 ± 0.019	1.24 ± 0.11
1/29/2010	129.82°	1.662 ± 0.004	2.253 ± 0.008	1.225 ± 0.008
3/3/2012	163.6665°	1.6931 ± 0.0005	2.331 ± 0.0010	1.230 ± 0.001
5/22/2016	241.7760°	1.531 ± 0.0013	1.906 ± 0.002	1.230 ± 0.003
27/7/2018	304.1995°	1.375 ± 0.002	1.551 ± 0.003	1.219 ± 0.005

Results for Jupiter:

Date	Ecliptic longitude at opposition	Distance from Sun at opposition, R_P, (a.u.)	Instantaneous period, T_P (years)	R_P^2/T_P
8/14/2009	322.066 °	5.069 ± 0.090	10.932 ± 0.287	2.35 ± 0.10
9/21/2010	358.3839°	4.951 ± 0.022	10.770 ± 0.076	2.276 ± 0.006
10/29/2011	35.2951°	4.957 ± 0.006	10.799 ± 0.017	2.276 ± 0.006
2/6/2015	137.6303°	5.328 ± 0.003	12.462 ± 0.017	2.278 ± 0.003
3/8/2016	168.2997°	5.425 ± 0.003	12.13 ± 0.010	2.279 ± 0.003
5/9/2018	228.3586°	5.408 ± 0.002	12.836 ± 0.008	2.279 ± 0.002

Results for Saturn:

Date	Ecliptic longitude at opposition	Distance from Sun at opposition, R_P, (a.u.)	Instantaneous period, T_P (years)	R_P^2/T_P
3/22/3010	181.2707°	9.566 ± 0.041	29.418 ± 0.217	3.112 ± 0.035
4/3/2011	193.8506°	9.715 ± 0.068	29.964 ± 0.384	3.150 ± 0.061
5,23/2015	241.6343°	9.977 ± 0.008	32.255 ± 0.054	3.085 ± 0.012
6/3/2016	253.1207°	10.043 ± 0.018	32.701 ± 0.128	3.086 ± 0.019
6/27/2018	275.8580°	10.063 ± 0.009	32.767 ± 0.054	3.090 ± 0.007
6/20/2020	298.6580°	9.977 ± 0.024	32.329 ± 0.123	3.079 ± 0.019

Distance measurements

Results for Uranus:

Date	Ecliptic longitude at opposition	Distance from Sun at opposition, R_P, (a.u.)	Instantaneous period, T_P (years)	R_P^2/T_P
9/21/2010	358.606°	20.063 ± 0.172	92.928 ± 2.002	4.33 ± 0.12
9/26/2011	2.5937°	20.007 ± 0.117	93.030 ± 1.263	4.302 ± 0.077
10/15/2016	22.5111°	19.876 ± 0.046	90.991 ± 0.447	4.342 ±0.029
10/24/2018	30.5646°	19.842 ± 0.041	90.260 ± 0.397	4.362 ± 0.026

I can take the values of R_P^2 / T_P that come out of the fit, and combine them with the average values of the period, to get the average distance of the planet from the Sun, essentially the semi-major axis. (OK. Mathematically this is not exact. The average of R_P^2 / T_P is not the same thing as the average of R_P squared divided by the average of T_P, but it is surely close. I improve on this in Chapter 18.) The results are

Planet	Average distance from Sun (a.u.)
Mars	1.524 ±0.004
Jupiter	5.21 ± 0.01
Saturn	9.72 ± 0.16
Uranus	19.5 ± 0.1

The uncertainties are statistical and do not allow for the approximation made. There is clearly some information on the eccentricities of the orbits of the planets contained in the difference between the average and instantaneous values of T_P. The most obvious lies in the results for Jupiter. The first three values for the distance of Jupiter from the Sun were all less than the average value. The practical difficulty is that the sidereal period of Jupiter is almost 12 years, so that the planet has not moved very far from one opposition to the next. That is why I waited until 2015 to make any further measurements. The distance value that I obtained then was satisfactorily greater than the average value. Clearly, Jupiter was close to perihelion in the period 2009 – 2011 and is getting towards aphelion in 2015. I expand on this idea in the next chapter. For Saturn and Uranus I probably need to wait a few years to give them time to move through a greater fraction of their orbits. As I said earlier, I shall never run out of projects!

Appendix: Retrograde Motion

The aim here is to relate the shape of the retrograde loop of a superior planet to the parameters of its orbit. This is another of the more mathematical parts of these pages. If the ecliptic latitude is plotted against the longitude, typical loops have shapes like these:

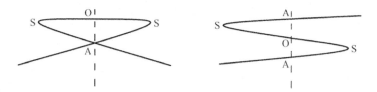

Figure 14.6 Shapes of retrograde loops

O is the opposition point and the S's are the stationary points. The first simplification I make is to assume that the planet moves in exactly the plane of the ecliptic i.e. in the same plane as the Earth moving around the Sun. The diagrams are then scrunched into horizontal zig-zag lines. In a time interval t the Earth revolves through an angle α_E and the planet moves through an angle α_P. Choose to measure time from the instant of opposition; I need only differences in times so I can make this choice. A picture after time t is now:

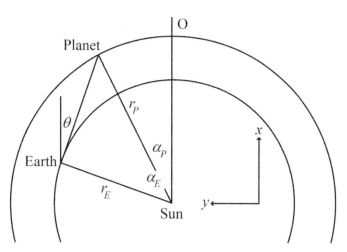

Figure 14.7 Geometry of retrograde motion

The planet started out at position O with the Earth directly below it. After time t the planet and the Earth have reached positions shown. The angle

Distance measurements

of sight looking from the Earth to the planet and the stars beyond it is labeled θ. One way to find this is to choose a pair of x and y axes, as I have, and to write components of the positions of the Earth and the planet.

$$x_E = r_E \cos(\alpha_E),$$
$$y_E = r_E \sin(\alpha_E),$$
$$x_P = r_P \cos(\alpha_P),$$
$$y_P = r_P \sin(\alpha_P),$$

and now find the angle θ by

$$\tan(\theta) = \frac{y_P - y_E}{x_P - x_E} = \frac{r_P \sin(\alpha_P) - r_E \sin(\alpha_E)}{r_P \cos(\alpha_P) - r_E \cos(\alpha_E)}.$$

At opposition, both the angles α_P and α_E are zero, and their sines are also zero, and hence θ is zero, as it must be. But it is also possible for θ to be zero if the two terms in the numerator are equal and subtract to give zero. This corresponds to the points marked A in the top diagram. They are the points where the value of ecliptic longitude of the planet is equal to its value at opposition. These are easy to locate accurately on a graph. The equation to determine them is

$$r_P \sin(\alpha_P) = r_E \sin(\alpha_E).$$

You can simplify this by assuming that the Earth and the planet both move in circular orbits with constant angular speeds, ω_E for the Earth and ω_P for the planet. In that case, after a time t the Earth and the planet have moved through angles $\omega_E t$ and $\omega_P t$ respectively. If you can measure the time difference t_A between either of the A points and opposition, that time is related to the distance of the planet by

$$r_P \sin(\omega_P t_A) = r_E \sin(\omega_E t_A)$$
$$\therefore \frac{r_P}{r_E} = \frac{\sin(\omega_E t_A)}{\sin(\omega_P t_A)}.$$

This is quite convenient to use. The angular speed of the Earth is $2\pi/T_E$, where T_E is the period of the Earth, or 1 year and the radius of its orbit is 1 astronomical unit. The factor of 2π gives the angle in radians. If you are using a hand calculator, you might need the angles in degrees and the 2π should be replaced with $360°$. Let's agree to measure periods in years and distances in astronomical units. Then the angular speed of the Earth is just 2π. If you have measured the sidereal period of the planet as T_P years, then the radius of the orbit of the planet is

$$r_P = \frac{\sin(2\pi t_A)}{\sin(2\pi t_A / T_P)} \quad \text{in a.u.}$$

The next simplest quantity to measure is the angular speed of the planet at opposition, as seen from the Earth. Sidney Brewer used a version of this method in his book. It has the advantage that you don't need to measure the complete curve. You measure the position of the planet for a few days around opposition, and the angle it moves through in that time, divided by the time period, is the angular speed. I shall call it Ω (upper case omega). The math for this gets a bit harder because I need some calculus. In terms of θ, the angular speed is a derivative with respect to time. The full expression for it is quite complicated but I need only two bits of the result. First, as I just indicated, I need the value at opposition, which turns out to be

$$\text{At opposition } \Omega = \frac{d\theta}{dt} = \frac{\omega_P r_P - \omega_E r_E}{r_P - r_E}.$$

The two ω's are 2π and $2\pi/T_P$ as before. The equation can be re-arranged as

$$r_P = \frac{2\pi - \Omega}{2\pi / T_P - \Omega} \quad \text{in a.u.}$$

The things to watch here are the odd units for the angular speed, which needs to be radians per year, and also that the planet is moving backwards so the measured speed should be negative.

The remaining numbers I can identify are connected with the stationary points, marked S. The easiest calculation uses the angle between opposition and one of the stationary points and the time taken by the planet to go from O to S. Call these θ_{st} and t_{st}. I can plug these values into the general equation for $\tan(\theta)$ to get

$$\tan(\theta_{st}) = \frac{r_P \sin(\omega_P t_{st}) - r_E \sin(\omega_E t_{st})}{r_P \cos(\omega_P t_{st}) - r_E \cos(\omega_E t_{st})}.$$

Putting in all the values I used before, and solving for r_P

Distance measurements

$$r_P = \frac{r_E \{\tan(\theta_{st})\cos(\omega_E t_{st}) - \sin(\omega_E t_{st})\}}{\{\tan(\theta_{st})\cos(\omega_P t_{st}) - \sin(\omega_P t_{st})\}},$$

where ω_E is 2π and ω_P is $2\pi/T_P$. If you use the average value of 1 for the distance of the Earth,

$$r_P = \frac{\tan(\theta_{st})\cos(\omega_E t_{st}) - \sin(\omega_E t_{st})}{\tan(\theta_{st})\cos(\omega_P t_{st}) - \sin(\omega_P t_{st})} \text{ in a.u.}$$

This is a little fussy, but you if you have all the numbers it can be solved easily in a spreadsheet or a short Basic program. These last two formulas turn out to be more sensitive to the quality of the data because, in numerical analysis terms, they involve numerical differentiation, which is a notoriously unreliable process.

In the chapter, I also used measurements of the quadrature point, when the line joining the superior planet to the Earth makes a right angle with the line joining the Earth to the Sun. The picture is the same as Figure 13.1, except that the positions of the planet and the Earth are reversed. For the quadrature point coming after opposition the formula is

$$r_P = \frac{r_E}{\cos(\alpha_E - \alpha_P)}.$$

Chapter 15: Mars, Jupiter, and Vesta

As I pointed out earlier, Saturn and Uranus move with sidereal periods longer than I have been making measurements for. While I shall keep tracking them, I am not expecting any surprises. My results for Mars and Jupiter have each given me new insights into their orbits.

Mars has turned out to present the most problems. I tracked the retrograde path of Mars very fully in 2016. I actually followed it from early March to the beginning of October. I put the data into the same program that I had used for Jupiter but the fit was very poor. Initially I abandoned the fitting idea completely. Using data from the Observer's Handbook I wrote a program that used elliptical orbits for both planets. This did fit the measurements very well. At this point I recognized that the time interval I was trying to fit was almost a third of Mars' sidereal period, so it was optimistic to represent that part of its orbit by an arc of a circle, which is what my fitting program did. I also looked back at the fits to the earlier oppositions of Mars. The quality of the fits was not quite as high as the fits for Jupiter, although I had been prepared to accept them at the time. I also realized that, by chance, at the earlier oppositions Mars had been moving through aphelion, when its angular velocity and distance from the Sun would have been changing relatively slowly, whereas, at the 2016 opposition Mars was just about half way between aphelion and perihelion, so that its angular velocity and distance from the Sun would have been changing at their greatest rates. All of this indicated that I needed to modify the fitting program to let these changes happen. But I did want to keep the changes as simple as possible.

I added one new parameter, which measured the rate of change of the angular velocity. If you took an introductory physics course, you probably met up with the set of equations describing a point object moving with constant acceleration. I used the same equations applied to the angular acceleration. As for the distance from the Sun, Kepler's second law relates that to the angular velocity for an elliptical orbit, so I could use that relationship without adding another parameter. In the previous chapter I did claim to have verified this law experimentally. The resulting program gave good fits to the new results, and I also went back and reanalyzed the results from the earlier oppositions with better agreement than the earlier fits. The fitted curve for 2016 is on the next page. As for the curve for Jupiter 2015 in the previous chapter, the line is the output from the fitting program rather than a smooth line drawn by Excel.

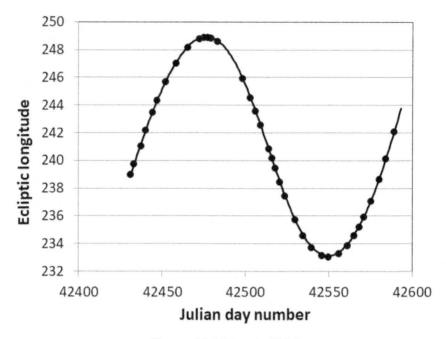

Figure 15.1 Mars in 2016

Just as I felt I was getting on top of Mars I had another setback. I followed Mars in 2018 from the very beginning of the year until the middle of October, although the retrograde motion occupied only the last part of this. The catch is that Mars went through perihelion, about eight weeks after opposition but still within the period I try to analyze. So Mars was speeding up for the first part of the period but slowing down for the last part. Neither my original program which treats the angular speed as constant nor the improved version which treats the angular acceleration as constant did a good job. I did run the data through the program that I first used for the 2016 data, that uses elliptical orbits for both planets. It gave excellent agreement with my measurements, not only for the retrograde loop but for the whole ten and a half months. Just to get a set of numbers for the previous chapter, I fitted just a narrow part of the data, leaving out both the beginning and the end. Even then, the fit was not quite up to the standard of the fits for other years.

If Mars brings me problems, Jupiter is a beautiful illustration of the elliptical orbits of the planets. The measurements that I made in 2009, 2010, and 2011 were grouped around the perihelion point and, by design on my part, the measurements for 2015, 2016, and 2018 were grouped around the aphelion point. I can plot the two sets of results separately,

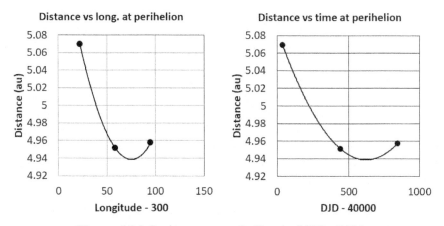

Figure 15.2 Jupiter near perihelion in 2009 - 2011

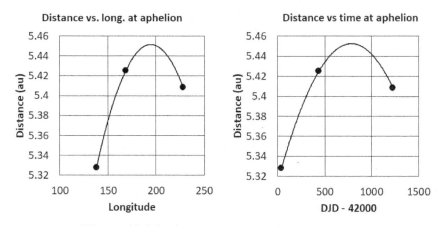

Figure 15.3 Jupiter near aphelion in 2015 - 2018

The lines are simple parabolas fitted by Excel, but a simple check on the adequacy of these fits is that the two curves around perihelion should give the same value for the perihelion distance and similarly the two curves around aphelion should agree on the aphelion distance. In both cases there is a small difference in the fourth decimal place between the two values. The averages of the two give a perihelion distance of 4.9385 a.u. and an aphelion distance of 5.4920 a.u. From these I can calculate a semi-major axis of 5.195 ± 0.005 a.u. and an eccentricity of 0.049 ± 0.001. I estimated the uncertainties by changing the values for the first perihelion point by one standard error, since this is the least precise of all the points. The times of the perihelion and aphelion points must differ by one half of a sidereal period and this gives a sidereal period 11.84 years for this particular cycle. A final value is that the longitude of Jupiter at perihelion is 14.6 ± 1.0°. I actually got this as 180° less than the aphelion angle, which is better defined. I am trying, at least up to Chapters 17 and 18, not to commit myself to an elliptical orbit picture, but these results make it difficult to avoid!

2018 turned out to be a very productive year. My first idea was just to follow Jupiter through its retrograde motion. This began in January and it turned out that Jupiter was in a very pretty close conjunction with Mars and with the star Zubenelgenubi, Since the first few photographs included Mars, I kept following Mars as the two planets moved apart. After a couple of months Mars came close to Saturn so I added it to the list. Then a club member pointed out that the asteroid Vesta was making an unusually bright appearance near Saturn. I checked and found that I already had twelve photographs of Vesta that I could analyze so I followed it through its retrograde motion also. The most exciting bit of this was the observation of Vesta. I had never tried to track an asteroid and didn't know what to expect. For the major planets I plot the ecliptic longitude as a function of time and fit a curve to extract the distance and period of the planet. But what would that plot look like for Vesta? Basically an asteroid is a minor planet and it has a typical planet's near-elliptical orbit. The plot is shown here.

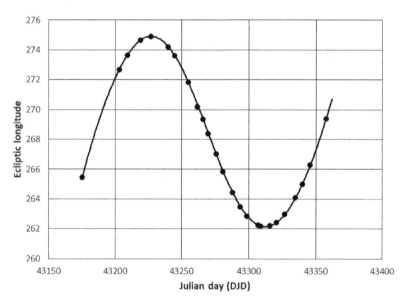

Figure 15.4 Retrograde motion of Vesta in 2018

As in the plot for Mars, the line comes from the fitting program, which also gives a value of 2.15 a.u. for the distance of Vesta from the Sun at opposition, and 3.04 years for the time it would take Vesta to complete an orbit if it kept going at a constant speed. Using some of the ideas that I develop in Chapter 18 I can estimate that its semi-major axis is 2.3 a.u. and that its sidereal period is about 3.5 years.

The photograph shows Vesta on 7/28/2018. Vesta is right in the center of the photograph, though it may not be easy to see!

Figure 15.5 Vesta on 7/28/2018

The four bright stars below it are c-, b-, θ- and o-Oph and the bright star above it and to the right is ξ-Oph. The magnitude of Vesta was 6.3. I just managed to see it in my telescope.

Chapter 16: Inclinations of the Planetary and Lunar Orbits

In all my thinking up to this point, I have assumed that the solar system is basically flat, that the orbits of all the planets and the Moon lie in a plane. I knew that wasn't correct but I, like most people, find it easier to think about two-dimensional situations than three-dimensional ones. When I did feel that it was time to let the third dimension creep in, I realized that the measurements I had already made contained enough information for me to extract the inclinations of the orbits of the other planets. I have put the math in an appendix, but I hope that Figure 16.1 below conveys the essential idea.

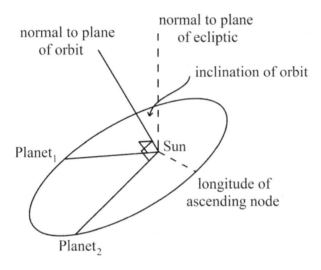

Figure 16.1 Inclination of a planetary or lunar orbit to the ecliptic

The plane of the ecliptic is taken to be horizontal. The oval is the elliptical orbit of a planet with the Sun at one focus, seen in perspective. I need to find the direction of the planet from the Sun at two separate times, when the planet's positions are Planet₁ and Planet₂. The two lines from the Sun to the planet lie in, and define, the plane of the planet's orbit. The line that is perpendicular to them is also perpendicular to the plane of the orbit. If I can write down expressions for these two lines as vectors, using the ecliptic coordinate system, then standard methods of vector algebra let me calculate the expression for the normal. The angle this line makes with the normal to the plane of the ecliptic is the angle through which the plane of the orbit is tilted. The component of the normal in the plane of the ecliptic contains information about the orbit that is usually expressed in the slightly archaic sounding longitude of the ascending node.

Distance measurements

In the case of Venus, I have described the measurements I made at its greatest eastern and western elongations in September 2010 and January 2011. I measured both the right ascension and declination at maximum elongation, when the line from the Earth to Venus makes an angle of 90° with the line from Venus to the Sun. That turns out to be enough information for me to construct the two vector lines, Sun - $Planet_1$ and Sun - $Planet_2$. For Mercury, I combined one of my maximum elongation measurements with a more recent observation of a transit. For each of the superior planets I have accurate measurements made for at least two oppositions. At opposition, the line from the Sun to the planet is in the same direction as the line from the Earth to the planet, expressed as the ecliptic longitude, except that there is a small out-of-plane component. I can calculate that component using the ecliptic latitude of the planet and the two distances of the Earth and the planet from the Sun. The expressions are all in the appendix.

I put in all the numbers and the results for the inclinations of the orbits came out much better than I had expected. The values of the longitude of the ascending node showed a little more variation from the handbook values, but were not too bad.

Planet	Inclination	Longitude of Ascending Node
Mercury	7.0°	49°
Venus	3.382°	74.8°
Mars	1.855°	45.3°
Jupiter	1.305°	99.6°
Saturn	2.490°	114.2°
Uranus	0.794°	68.5°

It is very difficult to estimate the uncertainties on these values because so many numbers have gone into the calculation. The least precise are the values for Mercury. If I use different combinations of observations the value of the inclination changes by ten percent. By changing the input values by plausible amounts, I find that for Venus the inclination is good to one or two hundredths of a degree and the longitude of the ascending node to perhaps three degrees. My other results are mostly in agreement with the handbook values to this sort of accuracy; the longitudes of the ascending node are off by a few degrees, up to 5° in the case of Uranus. If the normal to the orbit was exactly perpendicular to the plane of the ecliptic, there would be no ascending node and its longitude would be completely undefined. Since the tilts of the various orbits are small, I did expect the longitudes of the ascending nodes to be less well determined.

I have made an approximation in this analysis. In each case, the two measurements were, necessarily, made at different times. I have to assume that the orientation of the orbit has not changed in this time. For the planets, I think this is fine, but when I apply the same formulas to the Moon I am on more shaky ground. In fact, I began thinking about this set of ideas mainly in connection with the Moon. As I mentioned way back in chapter 1, I was foiled in several attempts to see new moons because the Moon was below the ecliptic, and that got me thinking.

Up to this point, I hadn't been successful using the method of photographing star fields to find the coordinates of the Moon. The Moon is so bright that an exposure that shows a useful number of stars overexposes the Moon so much that its image is hidden by massive amounts of flare. I hit on one way round this more or less by accident. I took a star field photograph that included the Moon at a time when the Moon was about two days from new, and its dark part was illuminated by Earthshine. I realized that I could use an exposure long enough to bring out the stars, and while the bright arc of the Moon was very over-exposed the "dark" part was quite bright and easily measured. My values are still not as accurate as the values for the planets. There is the extra step of locating the center of the image of the Moon, and also the photographs of a more-or-less-new Moon are taken when the sky is not quite dark, so there are fewer star images to work with. There is still another correction to be made. My observations are made from a point on the Earth's surface and the right ascension and declination values that I get are so-called topocentric values. I need the numbers that would be measured by an observer at the position of the center of the Earth. There is a program in Duffett-Smith's set of routines that makes the correction and I have relied on it. For the Moon, the correction is well outside the uncertainties of the measurements, and it must be made before the ecliptic longitude and latitude are calculated.

Initially I was interested in the inclination of the Moon's orbit to the ecliptic. I was concerned that the orbit wobbles in some way and I wanted to use pairs of observations that were as close in time as I could make them. I photographed the Moon a few days before new and a few days after new for seven successive months, and used each pair of ecliptic coordinates to find the inclination. The values I got ranged from 4.97° to 5.28°, with an average of 5.15°. This is very close to the handbook value of 5.16° but I really don't know if the variation between the values is experimental uncertainty or a slow change in the value. The difficulty with the Moon is that its orbit is so complicated that anything might be happening.

Distance measurements

When I had made this set of measurements, I realized that the numbers I was collecting would lead to a determination of another version of the lunar month, namely the draconic month. This measures the period with which the Moon bobs above and below the ecliptic. A plot of the ecliptic latitude as a function of time oscillates with this period. I really don't have measurements taken over a long enough period to give a precise value for this. Back in chapter 2, when I was talking about the synodic and sidereal months, I pointed out that there are oscillatory corrections to the length of each of them and it is necessary to average over a long period, perhaps ten years, to get a value correct in the second decimal place. If that is true in this case also, then my value might drift a little as I continue the measurements. However, there is nothing to stop me trying to pull out the best value that I can with the data I have. A fit of a sine function to my present results gives a period of 27.212 ± 0.002 days. I could see from the output of successive iterations of the least squares calculation that the value was slightly changing from one iteration to the next. The data are not good enough to pin down the minimum point precisely. When I first tried this my value was quite a bit below the handbook number and I was concerned that I was doing something wrong in the analysis. I fabricated a set of data points for the same period using the computer program MICA. The fit to those points gave almost exactly the same answer as the fit to my data points. So I assume that the Moon goes through periods of its motion when this version of the month is below or above average, and that in time its average settles down at the long-term value. The fascinating aspect of the result is that it is a bit *less* than a sidereal month, so that the line of nodes is rotating in the "wrong" direction compared with the advancing perigee.

I can get confirmation of that from the values of the longitude of the ascending node calculated from the successive pairs of values. I have now measured enough pairs of values to define this slope quite well. I can follow my usual method of plotting the values in Excel and asking for a linear trend line. The best value I can extract at present for the rate of decrease is 0.05344°/day, with a 1% uncertainty. This corresponds to a complete rotation through 360° in 6736 days or 18.4 ± 0.2 years. I can use that as a back door to estimate the long-time average value of the draconic month. The 18.4 year rotation period comes about from the difference between the sidereal and draconic months. I have a very accurate value for the long-term average of the sidereal month and I can make a small correction to that to find a value for the draconic month of 27.211 ± 0.001 days. This agrees well with the directly fitted value and with the handbook value. The graph is shown below.

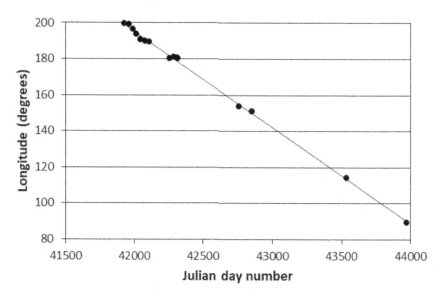

Figure 16.2 Longitude of the ascending node of the Moon

Jean Meeus shows a calculation of the sloping line as Figure 1.c in his book "Mathematical Astronomy Morsels". My results show small oscillations which are quite similar to the ones in Meeus's curve but I am not sure if mine are significant given the crudeness of the measurements

   I initially set up this calculation to find the ecliptic latitude of the Moon as a function of time, but as a bonus I get the ecliptic longitudes as well. The Observer's Handbook lists a value for the Tropical Month. This is almost identical with the sidereal month but measures the month relative to the ecliptic coordinate system rather than to the actual fixed stars. I can make a plot very similar to those for the synodic and sidereal months and extract a value for tropical month. There are two catches. The big one is that my data cover only five years while the other plots cover thirty years. The best value I can extract at the moment is 27.32 ± 0.01 days. The second catch is that I use several programs to process the results and I am not sure if part of the answer is built into them.

Distance measurements

Appendix: Calculating the orbit inclinations

For the planets, the first step is to find the heliocentric coordinates of the planets at two points in their orbits. For several special positions of each of the planets, I have found their right ascensions and declinations. From those, I can find their ecliptic longitudes and latitudes, with the help of Duffett-Smith's programs or the formulas in Meeus's book. Let me write the ecliptic longitude and latitude as a and b, and the heliocentric longitude and latitude as c and d. The relationship between the two pairs of numbers is different for the inferior and superior planets.

I described in chapter 6 how I could find the heliocentric longitude of Venus at two maximum elongations. It is equal to its ecliptic longitude plus or minus 90°, depending whether it is an eastern or western elongation. To find the heliocentric latitude, I need the distances of Venus from both the Earth and the Sun. Look back at Figure 13.1 from chapter 13.

The angle θ is the maximum eastern elongation of Venus from the Sun. The distance r_P from the Sun to the planet is $r_E \sin(\theta)$, and the distance from the Earth to the planet is $r_E \cos(\theta)$. If the ecliptic latitude of the planet is b, the linear distance of the planet above the plane of the ecliptic is this second distance multiplied by $\sin(b)$, and I can also express this distance in terms of the heliocentric latitude and the distance of the planet from the Sun. Setting the two expressions equal gives

$$r_P \sin(d) = r_E \cos(\theta)\sin(b)$$

$$\therefore \sin(d) = \frac{r_E \cos(\theta)\sin(b)}{r_P}$$

$$= \frac{\cos(\theta)\sin(b)}{\sin(\theta)}.$$

Once you have the sine, take the inverse sine to find the angle.

For the superior planets I use the values that I found at two successive oppositions. At opposition, the heliocentric longitude is equal to the ecliptic longitude, because the Sun, the Earth, and the planet are all in a straight line. The distance of the planet from the Earth is $(r_P - r_E)$, and the linear distance of the planet above the plane of the ecliptic is this distance times $\sin(b)$, as for the inferior planet. For r_P I use the instantaneous value that I found in chapter 14. The distance above the ecliptic plane is also equal to r_P times $\sin(d)$ and setting the two expressions equal gives

$$r_P \sin(d) = (r_P - r_E)\sin(b)$$
$$\therefore \sin(d) = (1 - r_E / r_P)\sin(b).$$

(If you have stuck with me this far, you might object that I seem to have used the value of r_P both for the total distance of the planet from the Sun and for its distance measured in the plane of the ecliptic. You would be correct. I have not been totally consistent. Fortunately, the values of the ecliptic and heliocentric latitude are so small that I can get away with this. I have done it both ways and the difference is completely insignificant. This way is the simplest to describe.)

Once you have the values of the angles c and d for two positions of the planet, the three-dimensional vectors connecting the planet to the Sun have components

$$x = \cos(c) \times \cos(d),$$
$$y = \sin(c) \times \cos(d),$$
$$z = \sin(d).$$

Put a subscript 1 on the components of the vector for the first position and 2 for the second position. The vector product of the two vectors is

$$y_1 z_2 - z_1 y_2, \; z_1 x_2 - x_1 z_2, \; y_1 z_2 - z_1 y_2,$$

which you might remember from high school. This points in the direction of the normal to the orbit. A convenient next step is to normalize this vector. Add the squares of the three components and take the square root of the sum. Divide each of the components by this square root and the vector is normalized (its length is 1). The z-component is now just equal to the cosine of the angle of inclination of the orbit.

In the case of the Moon, the ecliptic longitude and latitude are just what I need for the calculation. That is, replace c and d in the formulas with a and b in the formulas for x, y, and z. The only new complication is, as I mentioned in the chapter, that I need geocentric rather than topocentric values of the right ascension and declination to put into the calculation. Duffett-Smith has a subroutine that makes the correction.

Chapter 17: Did Copernicus and Kepler get it right?

I can use the numbers I have measured in chapters 4 through 7 and 13 and 14 to verify some of the revolutionary ideas propounded by Copernicus and by Kepler. Copernicus was responsible for the widespread acceptance of the idea that the Earth and all the planets rotate around the Sun. This idea had been floated many times before but after the work of Ptolemy and Aristotle, the Earth-centered model dominated for over 1400 years. Two old but very good histories of astronomy are "A History of Astronomy from Thales to Kepler" by Dreyer and "A History of Astronomy" by Pannekoek. Ptolemy and Aristotle championed the epicyclic theory of planetary motion. In the most basic form of this theory, we picture for each planet a circular orbit around the Earth. The planet does not actually move around this circle. Instead, the planet moves on a smaller circle, the epicycle, whose center does move around the original circle. Both motions are at constant angular speed. Each orbit is specified by two numbers. One is the time it takes the center of the epicycle to move around the Earth, and the other is the time it takes the planet to move around its epicycle. The two times are determined separately for each planet so as to reproduce the observed behavior. There is no built-in relationship between them.

In Copernicus's theory, the planets all move round the Sun, with their sidereal periods. The Earth also moves around the Sun, taking one sidereal year for each orbit. Because the Earth is a moving platform, the time it takes for each planet to complete an orbit as seen from the Earth is different from its sidereal period. Think first about a superior planet. It is moving slower than the Earth. Suppose that at some time the planet is in opposition, so that the Sun, the Earth, and the planet are all in a straight line. When will the planet next be in opposition? After a time t, the Earth has moved through t/P_E revolutions and the planet has moved through t/P_s revolutions. Here P_E is the sidereal period of the Earth and P_s is the sidereal period of the planet. There is the usual caveat that the angles turned through are mean values. The synodic period of the planet, P_y, is the value of t for which the Earth has moved through exactly one more revolution than the planet. As an equation this is

$$\frac{P_y}{P_E} = 1 + \frac{P_y}{P_s}.$$

I can rearrange that in several ways. One gives P_y in terms of P_s and the Earth year P_E:

Distance measurements

$$\frac{1}{P_y} = \frac{1}{P_E} - \frac{1}{P_s}.$$

I'll get back to this later. Another arrangement is

$$P_E = \frac{P_s P_y}{P_s + P_y}.$$

This is the one I shall use to test Copernicus. I have measured the two periods, P_s and P_y, quite separately for each of the superior planets. I rely mainly on my measurements of oppositions for both periods, but they depend on different parts of the data; the synodic periods came from the interval between oppositions while the sidereal periods came from the angle the planet moved through between oppositions as well as the length of time that an opposition took up. The distances from the Sun depend mainly on the angles that the planets seem to move through during the retrograde motion. If Copernicus is correct, these two numbers for each planet are related to the length of a sidereal year for the Earth. How do the numbers look?

Planet	Sidereal Period, P_s	Synodic Period, P_y	$P_s P_y/(P_s+P_y)$
Mars	690 ± 4	776 ± 3	365.3 ± 1
Jupiter	4350 ± 22	398.7 ± 0.6	365.2 ± 0.5
Saturn	11177 ± 365	377.8 ± 0.2	365.4 ± 0.4
Uranus	31996 ± 365	369.5 ± 0.4	365.3 ± 0.4

The values are all in days. The ratios all come out equal to each other and to the Earth's sidereal year within the observational uncertainties.

For the inferior planets, Mercury and Venus, the relationship is slightly different. The alignment that corresponds to opposition for a superior planet is now an inferior conjunction. The Sun, the planet, and the Earth are then in line. If they are exactly in line there is a transit of the planet in front of the Sun. More often, the planet passes slightly above or below the Sun. While it is possible to see the planet in either case, it would be difficult to record its position relative to the stars. The inferior planets move faster than the Earth, so in the time from one inferior conjunction to the next the planet has moved through one complete revolution more than the Earth. The equation is

$$\frac{P_y}{P_s} = 1 + \frac{P_y}{P_E}$$

$$\therefore \frac{1}{P_s} = \frac{1}{P_y} + \frac{1}{P_E}.$$

This can be re-arranged as

$$P_E = \frac{P_y P_s}{P_y - P_s}.$$

Again, I have separate values for both the sidereal and synodic periods of the two inferior planets so that I can check this relationship.

Planet	Sidereal Period, P_s	Synodic Period, P_y	$P_y P_s/(P_y - P_s)$
Mercury	88.0 ± 0.1 days	115.9 ± 0.3 days	366 ± 3 days
Venus	224.68 ± 0.03 days	584.0 ± 0.25 days	365.2 ± 0.1 days

Also again, Copernicus is vindicated!

Armed with these results, I can test a part of Kepler's ideas. Kepler spent years analyzing the planetary measurements of Tycho Brahe, mainly the measurements on Mars. He eventually formulated three laws of planetary motion.

1. The planets move in elliptical orbits with the Sun at one focus of the ellipse.
2. As a planet moves, its speed changes in such a way that the line joining it to the Sun sweeps out equal areas in equal times.
3. The distances, a, of the planets from the Sun (actually the semi-major axes of their orbits) are related to their periods, P, by

$$P^2 \propto a^3.$$

I did actually test the correctness of Kepler's second law when I found the best values for the average distances of the planets from the Sun, in chapter 14, but now I want to test the third law. It is very convenient to use Earth-years to measure the periods, and to express the sizes of the orbits in astronomical units. Then for the Earth, P and a both take the value 1 and the ratio P^2/a^3 is also 1. I can find this ratio for each of the planets.

Distance measurements

Planet	P (years)	a (a.u.)	P^2/a^3
Mercury	0.2409 ± 0.0003	0.38 ± 0.01	1.06 ± 0.08
Venus	0.61514 ± 0.00008	0.723 ± 0.004	1.001 ± 0.017
Mars	1.89 ± 0.01	1.525 ± 0.004	1.009 ± 0.013
Jupiter	11.91 ± 0.06	5.21 ± 0.01	1.003 ± 0.013
Saturn	30.6 ± 1.0	9.72 ± 0.16	1.02 ± 0.08
Uranus	87.6 ± 1.0	19.5 ± 0.1	1.03 ± 0.03

The ratios are all equal to 1 within the uncertainties.

Another confirmation that the Earth goes round the Sun comes from the relationship between the synodic and sidereal months. The equation relating them is in fact exactly the same as the one for the inferior planets. In this case you can imagine starting from a new moon position, when the Moon is exactly in line between the Sun and the Earth. After one sidereal month, the Moon has completed an orbit of the Earth but, because the Earth has moved, the Moon has still not reached the next new moon position. So in a time equal to the synodic month, T_y, the Earth turns through a fraction T_y/P_E of a revolution, where P_E is just the one-year period of the Earth. The Moon turns through T_y/T_s revolutions and this must be a complete revolution plus the fraction turned through by the Earth. This gives me the equation I used above for the inferior planets. Another argument that leads to the same answer is the long-piece-of-string picture that I used for the sidereal year in chapter 3. Imagine that the Moon is connected to the Earth by a long piece of string. Every sidereal month, the Moon goes round the Earth and leaves one turn of string. As the Earth moves round the Sun, it tends to unwind the string. In a complete year it unwinds one complete turn of the string so that there is one fewer turn (i.e. one fewer synodic month) in the year than there are sidereal months. This leads to exactly the same equation. This time I want to solve it explicitly for the Earth year, P_E:

$$P_E = \frac{T_y T_s}{T_y - T_s}.$$

Writing the equation this way shows that the calculated length of the year depends on the difference between the synodic and sidereal months. This is not a good thing. There is a loss of precision when two numbers of similar magnitudes are subtracted one from the other. I have measured each of the months to at least five significant figures, but when I subtract one from the other, the most significant figure disappears and I am left with only four figures. The result is

$$P_E = \frac{29.530 \times 27.321}{29.530 - 27.321}$$

$$= \frac{29.530 \times 27.321}{2.209}$$

$$= 365.2 \pm 0.2 \text{ days}$$

I estimated the uncertainty based on an uncertainty of .001 in the difference of the two months. As far as confirming that the Earth goes round the Sun, this is fine. The answer is one year within the uncertainty. I would really have liked to get one more significant figure in the final answer. It was important to me that the measurements of the two months and of the sidereal year were all independent, but if I accept that this formula is correct, I can combine my rather good value for the sidereal month with the value I have measured for the sidereal year, P_E, to give a value for the synodic month using the relationship in the form

$$\frac{1}{T_y} = \frac{1}{T_s} - \frac{1}{P_E}.$$

This gives a value for the synodic month of 29.5306 ± 0.0005 days.

I had thought of calling this chapter "Did Copernicus, Kepler, and Newton get it right?', but I didn't have the nerve. Of course Newton got it right. He always did. What I was thinking about was his use of the Moon to test his theory of gravitational attraction. In chapter 18, I shall use Newton's results relating the period and orbital radius for a satellite moving around a central mass. I shall use them there for the motion of the Earth around the Sun, but the same equation describes the motion of the Moon around the Earth. It can be arranged as

$$T^2 = \frac{4\pi^2 r^3}{GM_E}.$$

T is a sidereal month, M_E is the mass of the Earth and r is the distance from the center of the Earth to the center of the Moon. Newton didn't know either the mass of the Earth or the gravitational constant G. Fortunately, he could get rid of both of them by using the expression for the acceleration due to gravity at the surface of the Earth

Distance measurements

$$g = \frac{GM_E}{R_E^2} \qquad \therefore GM_E = gR_E^2.$$

Newton did have good values for the acceleration due to gravity, g, and the radius of the Earth, R_E. He could therefore write

$$T^2 = \frac{4\pi^2 r^3}{gR_E^2}.$$

I, also, have values of my own for the quantities on the right, albeit less accurate, so I can calculate the period of the moon as 28 ± 2 days. This agrees with my measurement of the sidereal month, within the uncertainty, but my values for r and R_E have large uncertainties. I derived my value of g from the results given in Appendix E, and it is very accurate. If I use handbook values for r and R_E the moon's period comes out as 27.5 days. This is slightly off the observed number but that difference can be because I have ignored the effect of the Sun. As I said, Newton was always right!

I could reverse my thinking here also. If I accept the correctness for this equation I can re-arrange it to find the distance to the moon. The solution is

$$\frac{r}{R_E} = \left(\frac{gT^2}{4\pi^2 R_E} \right)^{1/3}.$$

My values for g and T are very accurate, and the cube root on the right side makes less important the large uncertainty in my value for R_E. The final value for r/R_E is 59.0 ± 1.

Appendix: Newton's theory of Gravitational Attraction

According to Newton's theory of universal gravitational attraction, a small object of mass m at a distance r from a spherical object of mass M experiences a force

$$F = \frac{GMm}{r^2}.$$

This applies outside the spherical object and all the way down to its surface. G is a universal constant.

As the Earth moves around the Sun, it experiences a gravitational force given by the formula above, where r is now the distance from the Earth to the Sun, m is the mass of the Earth, and M is the mass of the Sun. If I approximate the Earth's orbit by a circle, its acceleration is the so-called centripetal acceleration, equal to its orbital velocity squared, divided by the radius of the orbit. Newton's famous second law of motion, $F = ma$, then gives

$$\frac{GMm}{r^2} = \frac{mv^2}{r}$$

$$\therefore \frac{M}{r} = \frac{v^2}{G}.$$

Notice that the mass of the Earth canceled. In one sense this is a nuisance; it means that I have no way to find the mass of the Moon, or of a moonless planet such as Venus. The speed v is equal to the length of its orbit, $2\pi r$, divided by the period of the orbit, T. So

$$v = \frac{2\pi r}{T}$$

$$\therefore \frac{M}{r} = \frac{1}{G}\left(\frac{2\pi}{T}\right)^2 r^2$$

$$\therefore \frac{M}{r^3} = \frac{1}{G}\left(\frac{2\pi}{T}\right)^2$$

$$\therefore T^2 = \frac{4\pi^2 r^3}{GM}.$$

I shall use this in chapter 20, to relate the mass of the Sun to the sidereal year and the astronomical unit, but the same equation can be applied to the orbit

Distance measurements

of the Moon round the Earth, with T now being a sidereal month and r being the distance from the Earth to the Moon.

To apply it to the Earth, it is helpful to have an extra application of the same ideas. A small object near the Earth's surface falls with an acceleration g, which implies a force mg acting on it. Setting M equal to the mass of the Earth, M_E, r equal to the radius R_E of the Earth, and putting the two forces equal

$$\frac{GM_E m}{R_E^2} = mg$$

$$\therefore M_E = \frac{gR_E^2}{G}.$$

The mass of the small object canceled. The quantities on the right side are the acceleration due to gravity, g, the radius of the Earth, R_E, and the constant G. I shall use this equation also in chapter 20.

Chapter 18: Elliptical Orbits

For a moment, let me assume that the planetary orbits exactly satisfy Kepler's laws. I know that cannot be correct, but a pessimistic view of my measurements is that they are not precise enough to show any deviations.

If I accept the correctness of the ideas of both Copernicus and Kepler, there is only one independent value in the three quantities of the sidereal period, the synodic period, and the radius of the orbit. I can use this to optimize my values. The most precise value I have for Mercury is the sidereal period of 0.2409 ± 0.0003 yrs. From this, using the equation $P^2 = a^3$, I calculate an average orbit radius of 0.3872 ± 0.0003 a.u. The synodic period calculated from the sidereal period and the Earth's sidereal year remains at 115.9 days, but the uncertainty is slightly reduced to ± 0.2 days, so these would be my optimized, or "best" values. I can pull out one more number, or at least an estimate of it. The closest distance of Mercury from the Sun at a maximum elongation that I have measured is 0.307 a.u. in September 2016. If I assume that this is the perihelion distance and is equal to $a(1 - e)$, I calculate a value for the eccentricity of 0.207 ± 0.005. This assumption also implies that the longitude of perihelion is about 78° because that was the heliocentric longitude of Mercury at that elongation.

For Venus the sidereal period is currently my best value, at 0.61514 ± 0.00008 years, or 224.68 ± 0.03 days. From that I can find a synodic period of 583.81 ± 0.08 days and a semi-major axis of 0.72330 ± 0.00006 a.u. The directly measured values were 584.0 ± 0.25 days for the synodic period and 0.723 ± 0.004 a.u. for the orbit radius. They are consistent but less precise. I can make a more elaborate extra calculation for the eccentricity than I did for Mercury. An elliptical orbit is described by three parameters: the semi-major axis, a, the eccentricity, e, and the longitude of perihelion. I have good determinations of the distance and heliocentric longitude for seven maximum elongations, and I have a very precise value for the semi-major axis. The idea is to perform a least-squares fit of the other two parameters to the data. My first try was not a success! I started the least-squares process at a rather arbitrary set of values. The least-squares didn't converge well and the answers drifted as I carried out more iterations. So I went more cautiously. I set up a grid of values of the eccentricity and longitude of perihelion in TrueBasic and calculated the quality of fit parameter, χ, for each grid point. By gradually shrinking the spacing of the grid points I could pick out a pair of values that I knew were close to a best fit. I started the least-squares process from these values and it quickly converged to give a value of 0.0066 ± 0.0002 for the eccentricity and $(126 \pm 2)°$ for the longitude of perihelion. The only remaining properties of Mercury and Venus that I

Distance measurements

might be able to measure are their size and mass. I tackle the first of these in Chapter 21 on the angular sizes of the Sun, Moon and Planets, and I make a guess at the masses there by assuming that Mercury and Venus have the same density as the Earth.

For Jupiter, I came up with three ways to find a set of elliptical coordinates. The first was the plots shown in figures 15.2 and 15.3 of Chapter 15. They really are easiest to interpret in terms of an ellipse. They gave the values $a = 5.195 \pm 0.005$ a.u., $e = 0.049 \pm 0.001$, the longitude at perihelion of $14.6 \pm 1°$ and a sidereal period of 11.84 ± 0.02 yrs. The second way is an analysis similar to the method I used for Venus, but slightly more complicated because I was trying to fix all three parameters of the ellipse. I first set up in TrueBasic a three-dimensional grid of values for the semimajor axis, the eccentricity, and the longitude at perihelion and located a set of values which gave a pretty good fit to the numbers from the six oppositions I had measured. Then a non-linear least squares fit starting from this point gave me a best fit with parameters: $a = 5.1995 \pm 0.0004$ a.u., $e = 0.0490 \pm 0.0001$, and the longitude of perihelion equal to $14.9° \pm 0.15°$. The value of a implies, via Kepler's third law, a sidereal period of 11.856 ± 0.002 yrs, These are very similar to the values from Chapter 15.

In the appendix to the previous chapter I applied Newton's Law of Gravitational Attraction to an object moving in a circular orbit around a center of attraction such as the Sun. But it was perhaps Newton's greatest triumph to show that his equations lead more generally to elliptical orbits which satisfy Kepler's laws. In Chapter 14 I made an estimate of the average distances of the superior planets from the Sun using the values of the ratio R_P^2/T_P that came out of the fit to the retrograde motion of each planet. Here R_P is the distance of the planet from the Sun at the point of opposition and T_P is the time it would take the planet to complete an orbit of the Sun moving at a constant speed. In Chapter 14 I argued that this number must be close to a^2/P, where a is the semimajor axis and P is the sidereal period of the planet. If I assume the orbits are exact ellipses I can do better than this. In fact

$$\frac{R_P^2}{T_P} = \frac{a^2\sqrt{1-e^2}}{P}.$$

Combining this with Kepler's third law, $P^2 = a^3$, gives a convenient alternate form

$$\frac{R_P^2}{T_P} = \sqrt{a(1-e^2)}.$$

The sidereal period is then given by $P^2 = a^3$ as usual.

For a third approach for Jupiter, I first find the best average value of the numbers for R_P^2/T_P that were listed in Chapter 14. I use that in the second form of the equation, taking the value of e from the first method as 0.049 ± 0.001, to find a value for a of 5.203 ± 0.008 a.u. and this would imply a sidereal period of 11.87 ± 0.03 yrs. The table below has the most precise of all these values.

For the other planets, the catch is that I don't have values for the eccentricity e. I do have fairly good values for either the perihelion distance or the aphelion distance for each of the planets. For Mars, Saturn and Uranus this seems to be the best route. I first calculate the best average values for R_P^2/T_P. Then, in the second form of the equation I first assume that e is zero and calculate a value for a. For Saturn and Uranus I then assume that the largest value I have measured for R_P is close enough to the aphelion distance, and use it to find a first approximation to e. For Mars, I assume that the smallest distance I have measured is the perihelion distance and this gives me a first estimate of e. I now put this back in the equation to find a better value of a, and keep repeating this series of steps until the numbers stop changing. For example, for Saturn the average value of R_P^2/T_P is 3.0898. If I use this to calculate a setting $e = 0$ I get a value of 9.5468 a.u. The largest value I have measured for R_P is 10.0631 a.u. so I use this as the aphelion distance. This lets me calculate an eccentricity of 0.05441. Putting this back into the equation and recalculating a I get a value of 9.5748, and I keep repeating these steps until the answers stop changing to a satisfactory degree. I actually did this in a TrueBasic program, but the convergence is quite fast. Only three or four iterations are needed. The final result for Saturn is $a = 9.56 \pm 0.03$ a.u. and $P = 29.57 \pm 0.14$ yrs, with $e = 0.052 \pm 0.003$. For Uranus, the average value of R_P^2/T_P is 4.349 ± 0.019. This would lead directly to an a value of 18.91 and a period of 82.28 years. Going through the iteration process gives final values of $a = 19.0 \pm 0.2$ a.u., $P = 83 \pm 1$ yrs, and $e = 0.057 \pm 0.013$. Since all my observations of both of these planets have been concentrated around the aphelion points these are by far the most realistic numbers that I have obtained, particularly for the sidereal periods. For Mars, the average value of R_P^2/T_P is 1.2291 ± 0.0008. This gives final values of 1.526 ± 0.002 for a, and 1.885 ± 0.004 yrs for P. The main weakness of the method is that the eccentricity depends on the difference between two similar numbers so it has a large relative uncertainty. That is why I didn't use this approach for Jupiter. I had a good value for e in that case from the other approaches.

Distance measurements

The values are all collected in the table below. The sidereal periods are not shown but they can be calculated from Kepler's third law, The best values for Mercury, Venus, Mars, Saturn and Uranus are given in the text. For Jupiter the most precise of my three estimates is 11.856 ± 0.002 yrs. Combine these results with the table in Chapter 16 of the inclinatio0ns of the orbit and the longitudes of the ascending node for each planet and with the measurements of the planetary sizes in Chapter 21 for a complete set of numbers. The masses of Jupiter and Saturn are tackled in Chapter 22.

Planet	Semi-major axis(a.u.)	Eccentricity	Longitude at perihelion
Mercury	0.3872 ± 0.0003	0.207 ± 0.005	78°
Venus	0.72330 ± 0.00006	0.0066 ± 0.0002	$126 \pm 2°$
Mars	1.526 ± 0.002	0.099 ± 0.002	344°
Jupiter	5.1995 ± 0.0004	0.0490 ± 0.0001	$14.9 \pm 0.2°$
Saturn	9.56 ± 0.03	0.052 ± 0.003	96°
Uranus	19.0 ± 0.2	0.057 ± 0.013	179°

The longitudes at perihelion for Mars, Saturn, and Uranus are simply 180° away from the longitudes of the measured points with the largest R_P values given in Chapter 14. They should be accurate to within a few degrees.

These values are probably the closest to the handbook numbers that I have obtained. But remember, they do depend on the assumption of elliptical orbits.

124

Part III: Physical properties

Chapter 19: One to get started - Surface temperature of the Sun

This is a bit of pure nineteenth century physics. My former colleague Frank (Casey) Blood pointed out to me the possibility of doing this. It involves the angular size of the Sun, which I shall also use in the calculation of the density of the Sun, and one more quantity, the solar constant.

The solar constant is the amount of solar energy, i.e. sunlight, per unit time, crossing a unit area surface at the distance of the Earth from the Sun. You can get a rough measurement of it by setting out a tray of water on a sunny, cloudless day, and measuring how quickly its temperature rises. From this humble experiment, you can find the temperature of the Sun.

I used a black 8" × 10" photographic tray to hold the water, and I put it on thin supports rather than directly on the ground, so as to minimize heat loss. For the same reason, I started the measurement with the water temperature several degrees below ambient and kept going until it was the same number of degrees above the surroundings, so that it would be gaining heat for part of the time and losing it for the rest. If the temperature of the water goes up by an amount ΔT, the amount of heat it has absorbed is $C \, \Delta T$, where C is the heat capacity of the water. (The upper case Greek delta, or Δ, notation is a common way to write the change in some quantity. ΔT is the change in the temperature T.) The heat capacity is proportional to the mass, m, of the water: $C = cm$, where c is called the specific heat of the water.

The actual experimental result was that the temperature of the water rose from 79.5°F to 89.5°F in 13 minutes. I also noted that the elevation of the Sun was 67°, and that I used two liters of water.

The experiment is simple and the analysis is simple in concept, but several detailed corrections are needed. From the measurements I need to calculate the rate at which the water absorbs energy from the Sun. I relate that to the surface temperature of the Sun using a formula called Stefan's law which describes how a hot object radiates. The volume of the water was 2 liters, or 2000 ml, so its mass was 2000 grams. In the nineteenth century the usual unit of heat was the calorie, defined so that the specific heat of water was just 1 calorie per gram/C°. The heat capacity of the water was therefore 2000 calories/C°. I have to correct this by adding on the heat capacity of the tray. I weighed it, and its mass was 283.5 g. It was made of an unknown plastic so I looked up the specific heats of various plastics and guessed that a value of 0.3 calories per gram-C° would be about average. The heat capacity of the tray then works out to be about 85 calories/C°. This is quite a small correction so I feel I know it accurately enough.

The increase in temperature was 10°F or 5.556°C. The heating rate of the water plus tray was then $2085 \times 5.556 / (13 \times 60) = 14.85$ calories per sec. The modern unit for this is the Watt, and 1 calorie per sec is 4.184 Watts, so the heating rate was 62.14 Watts.

The surface area of the top of the water was 768 cm². There is another correction here because the sunlight is coming at an angle. I have to multiply the area by the sine of 67°. There is a second correction because the front of the tray was also exposed to sunlight. It had an area of 92 cm² and the angular correction for that is cos(67°) because it is a vertical surface. Adding up all of these factors gave me an effective area of 742.9 cm² or 0.07429 m².

Finally, the solar constant is the power per unit area, which is 62.14W/0.07429m² or 836 W/m². This isn't a particularly good answer. The value given in handbooks is 1370 W/m². The main reason why my value would be low is that the handbook value corrects for absorption by the atmosphere whereas I am assuming that all the radiation from the Sun reaches the surface of the Earth. In the nineteenth century, when this was really "hot" physics, the experimenters would go to mountain tops to reduce the effect of the atmosphere. I didn't do that, and I actually had the optimism to do it in New Jersey, where even the bluest of skies has a lot of water vapor in it. So I am satisfied with the result. Fortunately, the number I actually want turns out to depend rather weakly on my answer.

Now I need to relate this value of the solar constant to the surface temperature of the Sun. The theory is in an appendix. The expression for the solar constant is

$$\text{solar constant } C_S = \sigma_B \left(\frac{r_S}{r}\right)^2 T_S^4,$$

where r_S is the radius of the Sun, r is the distance of the Sun from the Earth, and σ_B is a constant, the Stefan-Boltzmann constant, that can be measured in the laboratory. I have not measured either r_S or r, but I have measured the angular size, α, of the Sun, which involves their ratio. In radians

$$\alpha = \frac{2r_S}{r}.$$

The final result for the surface temperature of the Sun is

$$T_S = \sqrt[4]{\frac{C_S}{\sigma_B}\left(\frac{2}{\alpha}\right)^2}.$$

The handbook value for σ_B is 5.6704×10^{-8} W/(m^{2}K^4). My best value for α, from chapter 19, is 9.284×10^{-3} radians. Putting all the numbers in, I find that the surface temperature of the Sun is 5114 K. (The K here is the unit of temperature on the thermodynamic or Kelvin scale. It is the same size as the Celsius degree but the thermodynamic temperature has an extra 273 added to it.) As in other results of mine, not all of those figures mean very much. In fact, my value for the solar constant was about 40% low, so the value for the temperature, which involves the fourth root, is low by about 10%. A fair statement of my result is to round it to 5100 K and note that it is a lower limit. I am happy with that.

Physical properties

Appendix: Stefan's Law

I need to relate the solar constant to the surface temperature of the Sun. Again, all of this is nineteenth century physics. The Sun behaves as a so-called black body radiator. The rate at which each unit area of its surface radiates energy is proportional to the fourth power of its temperature on the absolute or thermodynamic scale. This is called Stefan's Law. The constant of proportionality is written σ_B and is called the Stefan-Boltzmann constant. To get the total rate of energy radiation from the Sun you multiply this by the surface area of the Sun, which is $4\pi r_S^2$ to get

$$4\pi r_S^2 \sigma_B T_S^4.$$

This energy is spread out in all directions and, at the distance of the Earth, it is spread over a sphere of radius r and surface area $4\pi r^2$. I divide the previous answer by this to get the solar constant

$$\text{solar constant } C_S = \frac{4\pi r_S^2 \sigma_B T_S^4}{4\pi r^2} = \sigma_B \left(\frac{r_S}{r}\right)^2 T_S^4.$$

This can be solved for the surface temperature T_S of the Sun

$$T_S = \sqrt[4]{\frac{C_S}{\sigma_B}\left(\frac{r}{r_S}\right)^2}.$$

The ratio of the radius of the Sun divided by its distance from the Earth, r_S/r, is one half of the angular size of the Sun as seen from the Earth, which I shall call α. The important point is that I know that quite accurately even though I don't know the distance from the Earth to the Sun or the size of the Sun. The final formula is therefore

$$T_S = \sqrt[4]{\frac{C_S}{\sigma_B}\left(\frac{2}{\alpha}\right)^2}.$$

I can't resist including a very neat calculation, even though it doesn't involve any more measurements of mine. The solar constant is the power per unit area passing through a surface at the distance of the Earth from the Sun. Seen from the Sun, the Earth is a disk of area πR_E^2, so the total power absorbed from the Sun by the Earth is

$$\pi R_E^2 C_S = \pi R_E^2 \sigma_B \left(\frac{r_S}{r}\right)^2 T_S^4.$$

What happens to this power? If the Earth is in equilibrium, the power must all be radiated away. The simplest assumption is that the Earth also radiates like a black body and the rate at which it radiates energy is then the same expression that I used for the Sun. In this case it is

$$4\pi R_E^2 \sigma_B T_E^4.$$

Setting these two expressions equal and solving for the temperature of the Sun gives

$$T_S = T_E \sqrt{\frac{2r}{r_S}} = \frac{2T_E}{\sqrt{\alpha}}.$$

α is, as before, the angular size of the Sun as seen from the Earth. My best value was 9.284×10^{-3} radians. Now, what do I use for the temperature of the Earth? In fact, I could guess it to sufficient accuracy! The temperatures in the formula are in Kelvins, which are bigger than the Celsius temperatures by 273, so I could the estimate an average temperature with sufficient accuracy. I have an outdoor thermometer that keeps a record of highs and lows, and in a typical year it varies between a few degrees and a hundred degrees Fahrenheit, so I might estimate that the overall average is somewhere around 55°F, or 13°C. Actually, because of the importance of global warming, average temperatures for the Earth do appear in the daily newspapers. A recently circulated value was 57° F or 14°C or 287 Kelvins. The calculated temperature of the Sun is then 5957 Kelvins, or close to 6000 K. This should be an overestimate because the Earth does not radiate as efficiently as a black body. My two estimates of 5100 K and 6000 K nicely bracket the handbook value.

Chapter 20: The Mass of the Earth and the Density of the Sun

Isaac Newton was the first person to be in a position to tackle this problem. In an appendix to chapter 17, I worked out this equation for the mass of the Earth

$$M_E = \frac{gR_E^2}{G}.$$

The quantities on the right side are the acceleration due to gravity, g, the radius of the Earth, R_E, and the constant G. Newton knew the values of the first two, but not the last one. What he actually did was first to estimate the average density of the Earth and this would let him find the other quantities.

I am more fortunate. I have measured R_E and also g. In fact the measured points in Appendix E: "The line goes through the points" can be analyzed to give a value for g of 9.80 m/s². I can look up the value of G in a handbook. It is 6.674×10^{-11} N-m²/kg². The mass of the Earth is then

$$M_E = \frac{9.80 \times (6750 \times 10^3)^2}{6.674 \times 10^{-11}} = 6.69 \times 10^{24} \text{ kg.}$$

Unfortunately, there is a 6% uncertainty in my value for the radius of the Earth, and when I square it this becomes a 12% uncertainty, so I should write my answer as $(6.7 \pm 0.8) \times 10^{24}$ kg.

A bit surprisingly, I can find the mean density of the Sun, using some more ideas from Newton's theory of gravity combined with a single extra measurement, namely the angular size of the Sun. This does need a bit more of the results in the appendix to chapter 17. There was an equation there relating the mass of the Sun and its distance from the Earth to the length of a year

$$\frac{M}{r^3} = \frac{1}{G}\left(\frac{2\pi}{T}\right)^2.$$

If I am prepared to look up the value of the astronomical unit, I can calculate the mass of the Sun, but I thought that it was neat that I can find the density of the Sun without knowing the value of the astronomical unit. The key idea is that this equation involves the mass of the Sun divided by the distance cubed. Now the density, ρ, of the Sun is its mass divided by its volume, and the volume of a sphere is $4\pi r_S^3/3$, where r_S is the radius of the Sun

Physical properties

$$\text{density } \rho = \frac{M}{V} = \frac{3M}{4\pi r_S^3}.$$

I need one more idea, which is that the angular size of the Sun as seen from the Earth, measured in radians, is its diameter divided by the distance from the Earth to the Sun

$$\text{angular size } \alpha = \frac{2r_S}{r}$$

$$\therefore r_S = \frac{\alpha r}{2}.$$

I used this idea in the last chapter also. Putting all these together, the density of the Sun is related to its angular size by

$$\rho = \frac{24\pi}{\alpha^3 G T^2}.$$

I have the value of G from the Observer's Handbook, T is just one sidereal year and I have measured that. It is 3.156×10^7 seconds. I measured the angular size of the Sun by taking photographs of it at prime focus of a 90 mm refractor at several times during a year. I describe that in chapter 21. My best value for the mean of the angular sizes is 31' 55" of arc, or 9.284×10^{-3} radians. Putting these values together I calculate a mean density for the Sun of 1420 kg/m³, or in a deprecated but familiar unit (1.42 ± 0.01) g/cm³.

If I do use the Handbook value for the astronomical unit, I can convert the angular size of the Sun into a linear radius using
$$r_S = \tfrac{1}{2}\alpha a_0$$
$$= \tfrac{1}{2}(9.284 \times 10^{-3} \text{ radians}) \times (1.496 \times 10^8 \text{ km})$$
$$= 6.944 \times 10^5 \text{ km}.$$
which is about 0.3% low. The diameter of the Sun is just twice this, or 1.389×10^6 km. The most direct way for me to find the mass of the Sun is to go back to the version of Newton's law that related the sidereal year to the mass, which I quoted earlier in this chapter. With r equal to 1 astronomical unit and G taken from the handbook, the only input from my own values is the sidereal year, which I have measured quite accurately. The value comes out as $(1.9885 \pm 0.0001) \times 10^{30}$ kg.

Chapter 21: Angular sizes of Sun, Moon, and Planets

The angular sizes of the various bits of the solar system are not exactly physical properties in themselves, but they turn out to contain quite a lot of extra information. This is the first set of measurements that really requires a telescope. For that very reason, the numbers that we get, at least for the planets, were not available to pre-telescopic observers. In fact, it was even worse for them, since they didn't realize the problem. Tycho Brahe thought that he could measure the angular sizes of stars as well as planets. What he was actually seeing was a diffraction spot whose size was determined by the opening in his eye. Galileo recognized that the images of the planets seen in his telescope were disks, but he still thought that he could measure the angular sizes of the stars. In fact, except when elaborate interferometer attachments are used on the largest telescopes, we still can't see the angular sizes of the stars. The disk images of planets become one of their distinguishing characteristics.

Of course, you don't need a telescope to see that the Sun and Moon are finite disks, and you can estimate their angular sizes using just the naked eye, with proper precautions to avoid looking directly at the Sun. But I wanted to get the sizes to around 1% accuracy or better, and to do that does need some optical help. I actually used two methods to measure the angular sizes, both of them rather crude in the twenty-first century. I initially tried a timing approach. The angular size of the Sun or a planet can be calculated from the time it takes to disappear from the field of view of a telescope, that is, the time from when it first touches the edge of the field of view to the time when it completely disappears. My initial tries at this were not very consistent and I moved on to a photographic method. More recently I gave the timing method a second try, and found that I could make it comparable in accuracy. I describe the details of the method in an appendix at the end of the chapter.

My main approach has been a simple photographic method. I described in chapter 10 how I calibrated my digital SLR mounted at the prime focus of a 90 mm aperture, 910 mm focal length refracting telescope. Using a white light solar filter, I took photographs of the Sun at several times of year, including around the equinoxes and at aphelion and perihelion. There is some fuzziness at the edges of the image of the Sun, which limits the precision of the measured diameters. John Rummel has given the suggestion, in the web page webpages.charter.net/darksky25/ Astronomy/Articles/sun/sunindex.html, to increase the contrast of the image by a substantial amount to give it a sharp edge. He was particularly interested in the difference in the diameters of the images at perihelion and aphelion.

Physical properties

The method is probably reliable for that, provided both images are treated consistently. I did find that I could change the size of the final images by adjusting the brightness before increasing the contrast.

My final numbers were based on measurements of both high contrast and normal contrast images, with what I hope is a fair estimate of the uncertainties. I photographed the Sun close to aphelion on July 8, 2009 and close to perihelion on January 4, 2010, and found angular sizes of 31' 25" and 32' 25" on the two dates. The midpoint of these should be the size at a distance of 1 astronomical unit, and it is 31' 55". I did also make a measurement close to the Fall solstice, on September 21, 2009, and found an angular size of 31' 47". Based solely on the consistency of the measurements, the uncertainties are about ±5", but making an allowance for the use of the high contrast technique, the absolute uncertainties might be twice that size. The diameter of the Sun, measured in astronomical units is just its angular size at a distance of one astronomical unit, expressed in radians. That comes to $(9.28 \pm 0.05) \times 10^{-3}$ a.u. or 1.389×10^6 km, using the handbook value for the astronomical unit.

The distance of the Sun at perihelion is $(1-e)$ a.u. and size at aphelion is $(1+e)$ a.u. where e is the eccentricity of the Earth's orbit, and the angular sizes are inversely proportional to the distances. The ratio of the sizes is 1.0322 ± 0.0033, and this corresponds to an eccentricity of 0.016 ± 0.002. This is not particularly precise, but the handbook value of 0.0167 is inside the limits.

I also used the timing method described in the appendix to measure the angular sizes of the Sun at perihelion, on January 4, 2014, and at aphelion on July 3, 2014. The two values for the angular size were 32' 27" ± 3" at perihelion and 31' 18" ± 14" at aphelion. The uncertainty for the aphelion value is larger because intermittent clouds cut down the number of measurements I could complete. The eccentricity of the Earth's orbit calculated from these values is 0.018 ± 0.008. The results are consistent with the numbers I got from the photographic method. But I do like the fact that the photographic method leaves a permanent record.

This all worked out quite well and I approached the Moon full of confidence. I expected to be taking a much greater number of photographs in this case. There are a lot of trees in my backyard and I thought it would be convenient to be able to move my position easily, so instead of the tripod mounted telescope I decided to use a hand-held 300 mm telephoto lens on the DSLR. The image size is smaller and the accuracy with which I can measure it is reduced but I thought the extra amount of data would compensate for that. Initially I concentrated on the full moon, partly because it is relatively easy to measure the size of the almost circular images, and I took photographs as close to the full moon as the weather would allow for three years. What would you expect the results to look like? My own reasoning was that if the Moon had an exactly elliptical orbit around the Earth, with a period of 27.3214 days, the image should oscillate up and down in size with this period and a sine curve should give a reasonable fit to it.

I used the same approach in chapter 16, for the draconic month. My thinking has a very simple component to it and a rather sophisticated one. The simple argument is what I just wrote. The measured points certainly oscillate up and down, and a sine function is the simplest built-in function that behaves this way. Furthermore, a circular orbit at a constant speed, viewed from the side, would look exactly like a sine wave. The more advanced argument lies in a pair of mathematical results associated with the mathematician Fourier. He showed that any truly periodic, or repetitive, behavior could be represented by a series of terms with the first term being the sine wave that I am using. The other terms oscillate at multiples of the basic frequency. If the quantity being modeled is not exactly periodic, it can be represented by an integral over sinusoidal terms with the basic sine curve at the middle of the integration. The final proof of the pudding is in the results. If the single sine curve does fit the points within the uncertainties, then it will be adequate. This is another reason for keeping track of the uncertainties. In this case, the results for my first three years of measurements are shown in Figure 21.1.

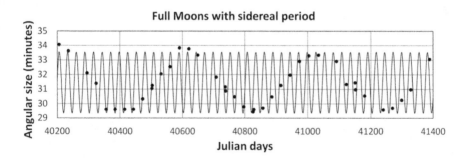

Figure 21.1 Angular size of full moons compared with sidereal period

Clearly there are points that lie above or below the sine curve. I was less concerned about them, because I really didn't think that a single sine curve would be an adequate approximation, as I just explained, than about the feature of the plot that the sine curve gradually gets out of step with the measurements, as you can see from the points at the right side of the graph.

Physical properties

It was fitted to just a few points at the beginning of the observation period. I have now continued the measurements for ten years and a least squares fit to the data gives a best fit period of 27.555 ± 0.001 days. I identify this with the so-called anomalistic month, which is the average time from perigee to perigee. The handbook value is 27.5546 days. The points at the right end of the set now fit the curve much better.

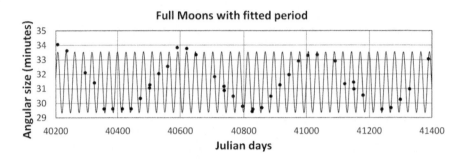

Figure 21.2 Angular size of full moon compared with fitted period

The difference between this period and the sidereal month, which I measured as 27.3214 days, arises because the perigee point of the Moon slowly advances; the elliptical orbit is precessing. My value for the anomalistic month corresponds for a period for the precession of 3237 days, or 8.86 years.

The formula for the fitted curve is

$$y = y_m + A \sin \left\{ \frac{2\pi (t - t_0)}{T} \right\},$$

where y stands for the angular size of the moon, y_m is its average value, T is the period of the oscillation and A is an amplitude, measured in arc-minutes. The term t_0 is adjusted to move the curve sideways along the axis. I initially adjusted the curve to fit the first few measured points. Actually, I did the original fitting after I had been collecting data for only one year, and measured an additional two years' worth of points to narrow down the value of T. I have now extended the measuring period by a further seven years' worth of measurements, which are not plotted here because the graph gets too crowded but which are included in my fit. I carried on the measurements because of my experience in measuring the synodic and sidereal months, that at least ten years' worth of observations would be needed for oscillatory terms to settle down. In fact, I am still collecting more data!

I wasn't completely happy about using the sine-function fit, because the points clearly are not going to be a really good fit to a sine curve. I did try a couple of other approaches to extracting the anomalistic month, such as using the MOD(,) function to reduce the points to a smooth curve for a single period, which I use very successfully in chapter 22 on Jupiter's moons, and I got very much the same value. I think I just need to go on collecting data.

If the orbit was indeed an ellipse, the amplitude A would give some indication of the eccentricity e of the orbit. Provided e is small enough that its square can be neglected, its value is just A / y_m. This comes out to be 6.7%. This is quite a bit bigger than the value given in the Observer's Handbook, but another handbook gave me a clue to that by saying that the standard value was an average, implying that the eccentricity varies with time.

At this point, I resolutely did not go to a book on celestial mechanics, which would presumably introduce me to the intricacies of the Moon's orbit. There is an often quoted comment of Isaac Newton's, in a letter to Edmund Halley, that his study of the Moon made his head ache! I figured that I should at least reach that point by myself. It seemed plausible to me that the numbers might well depend on the phase of the Moon. All of my measurements had been made at full moon, when the Sun, Earth, and Moon are more or less in a straight line, and the gravitational pulls of the Earth and Sun are adding together. I thought I might see some change in the numbers if I made my measurements at one-quarter and three-quarter moons, when the gravitational forces of the Earth and Sun were perpendicular to one another. So, while continuing the full-moon observations, I embarked on a similar series for the quarter and three-quarter moons. This did need a modification of my analysis technique, because the images are no longer complete circles but arcs. I wrote a short Basic program that accepts the coordinates of three points on the arc and finds the diameter of the circle they lie on. In practice, I select multiple sets of points and take an average. The results are in Figure 21.3. The calculated line has the same period and t_0 values as for the full moons, but the amplitude A and mean value y_m have been separately fitted. The value of the eccentricity calculated from these numbers is 4.4%. I would love to be able to find a value for it close to new moon, but I have very few photographs of the Moon close to new that can give me accurate values of its angular size. In his book "Mathematical Astronomical Morsels", Jean Meeus shows an amazing plot of the variation of the eccentricity of the Moon's orbit over a period of several years.

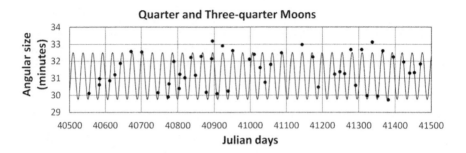

Figure 21.3 Angular size of quarter moon compared with fitted period

The fitted values of y_m, which are the mean values of the angular size of the Moon, are different in the two fits. For the fit to the full moon data the average is 31' 27" and for the quarter and three-quarter moons it is 31' 8". I can speculate on why they are different. One possibility is that the difference is caused by the differing directions of the attractive forces of the Earth and of the Sun. In the full-moon case, the two forces are in the same direction and it is plausible that this would cause the Moon to be closer to the Earth and hence to have a larger angular size. In the quarter-moon case, the two forces are at right angles. Unfortunately, I can come up with another, less interesting explanation. In chapter 12, on determining the distance to the Moon, and the full-moon effect in particular, I pointed out that the angular size of the Moon changes as the Moon rises in the sky. The full-moon photographs were all made with the Moon relatively high in the sky. The height it reaches depends on the time of year, but I did my best. The quarter- and three-quarter-moon photographs were sometimes made with the Moon somewhat lower in the sky, and this can account for some, perhaps most, of the difference. Perhaps I am beginning to get that headache!

I can pull one more number out of my results. The first set of results for the full moon had a sine curve already plotted on it. If I had left that off, you might have pictured for yourselves a sine curve with a much longer period as shown in Figure 21.4. The fitted period of this is 411.9 ± 0.2 days. I identify this with the so-called "super-moon period". Super-moons are exceptionally large full moons that occur when full moon coincides with perigee. The period is equal to $T_y T_a / \left(T_y - T_a \right)$, where T_y is

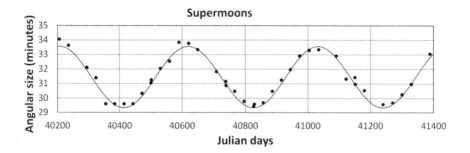

Figure 21.4 Angular size of full moon compared with supermoon period

the synodic month and T_a is the anomalistic month. Personally, I have found super-moons to be over-rated, but they do often get a small mention in the newspapers. I can use this result as a back door to what the long-term average of the anomalistic month will be, rather as I did for the draconic month in chapter 16. If the super-moon period is written as T_s, the relation between the three months can be rearranged as

$$\frac{1}{T_a} = \frac{1}{T_y} + \frac{1}{T_s},$$

My best value for the synodic month is already a long-term average, and the last term, containing the super-month period, is a small correction to this. This gives me a value for the anomalistic month of 27.555 also, so the "front-door" and "back-door" methods have converged.

This may be a good time to step back and try to visualize the orbit of the Moon. Start out with the idea that the orbit is an ellipse, though with quite a small eccentricity. In fact, as I have now verified, and Meeus illustrates beautifully, the eccentricity is not constant, which implies that the orbit is not exactly an ellipse. Now I find that the time it takes the Moon to go from one perigee to the next is slightly longer than a sidereal month. This means that when the Moon has completed a 360° path round the Earth, the perigee point has moved slightly ahead. The not-quite-an-ellipse is precessing in the same direction that the Moon is tracing its orbit. It takes about 8.86 years for it to complete a rotation. Now add into that picture the idea that the orbit of the Moon is slightly tilted with respect to the orbit of the Earth around the Sun. The planes of the two orbits intersect in a straight line. This line is also rotating, actually taking about 18.4 years to complete a rotation. But it rotates in the opposite direction to the advance of the perigee point. Personally, I give up the effort to visualize all of this.

My results do also illustrate another odd behavior of the Moon that is described by Jean Meeus in "Mathematical Astronomical Morsels". It is clear from Figures 21.2 and 21.3 that the angular size of the Moon deviates more from a simple sine curve above the curve than below it. The minima are almost constant. The angular size is greatest when the Moon is closest from the Earth, so these results mean that the perigee distance of the Moon from the Earth varies much more than the apogee distance. Figure 2.b in Meeus' book is a plot which shows just this behavior. The headache is still there!

For the planets, I again used the 90 mm telescope, this time supplemented with a photographic 3x extender, essentially the same as a 3x Barlow lens, to give a nominal focal length of 2730 mm. I calibrated the extender to find the exact multiplication factor and I did that by photographing the Moon with and without the extender and comparing the image sizes. Even with the extender, the images of the planets are not large. Jupiter at opposition and Venus as it approaches inferior conjunction give images of the order of 100 pixels diameter, Saturn and Mars are quite a bit smaller, 40 and 30 pixels respectively, and Mercury is wishful thinking!

Actually, I was able to get a value for the diameter of Mercury when I photographed its transit in front of the Sun in 2016, shown in Figure 4.1. At that point Mercury is at its closest approach to the Earth and its silhouette is a complete circle. I took that photograph without the extender in place in order to get the image of the whole Sun, but I also took a photograph using the extender. On the photograph the image of Mercury is 9 pixels across. This corresponds to an angle of 12.3" of arc. From my measurements of the orbit, Mercury is about 0.45 au from the Sun and therefore 0.55 au from the Earth. If it was moved away to be 1 au from the Earth its size would be 5 pixels. The image of the Sun is 1395 pixels across, so the size of Mercury divided by the size of the Sun is $5/1395 = 3.6 \times 10^{-3}$. My value for the diameter of the Sun is 1.389×10^6 kilometers, so the diameter of Mercury is just on 5000 km. The uncertainty in the size of the image of Mercury is at least ± 1 pixel and this translates into (0.0036 ± 0.0004) for the size of Mercury divided the size of the Sun and (5000 ± 600) km for the diameter of Mercury.

Venus was easily seen in the northern hemisphere sky as it approached inferior conjunction in December 2013 and January 2014. I took photographs of its thin crescent on several evenings. There were several potential difficulties. One was that all that the images showed was a partial arc of a circle rather than a complete disk. I had to use the same procedure that I had developed for the quarter and three-quarter moons. I picked off the coordinates of three points on the arc and then used the same TrueBasic program as before to find the radius of the circle. A more serious problem was that the arcs showed a lot of color fringing. I am sure that at least a part of this arose from the fact that Venus was necessarily low in the sky and I was photographing it through a curved atmosphere. I made what seemed the best choice and picked the points in the yellow band, in the middle of the fringes, but I bumped up the uncertainty estimate to try to cover myself. Finally, the distance to Venus changed substantially over the series of evenings. I took the distance from a planetarium program for each evening, and calculated the size of Venus for each case before taking an average. My final result was that the diameter of Venus is $(8.2 \pm 0.2) \times 10^{-5}$ a.u., or 12400 $\pm$ 300 km. This is 0.0088 $\pm$ 0.0002 times the diameter of the Sun. It is a little ironic that I am claiming to have measured the diameter of Venus with a smaller uncertainty than my measurement of the diameter of the Earth. That is partly due to my stubbornness in insisting that one of the reference points for my measurement of the Earth was in my backyard. Also, I would be very happy to be able to find the mass of Venus, but the only route I have to that would be to measure the orbit of a moon, and Venus has no natural moons!

I can make a guess at the masses of both Mercury and Venus if I assume that their densities are the same as the density of the Earth. I then find a mass of 3×10^{23} kg for Mercury and 5×10^{24} kg for Venus. These agree with the handbook values to one significant figure, but they are just a guess. If I use the same reasoning for Mars or the Moon, I get quite an overestimate.

I photographed Jupiter close to opposition on November 30, 2012. The images are not great compared with the results that the experts in my club can produce, but the line of the cloud bands gave the orientation and the oblateness was obvious. I made measurements both on the original images and on high contrast images. I found the equatorial diameter to be 102 $\pm$ 2 pixels and the polar diameter to be 92 $\pm$ 2 pixels. These translate into an angular size of 49" $\pm$ 1" equatorially, and an oblateness of $1/(12 \pm 4)$. The uncertainty on the oblateness is unfortunately large. To convert the angular size into an actual diameter, I need the distance. The diameter is just the distance multiplied by the angular size in radians. Based on my results from chapters 13 and 14, I can calculate the distances to Jupiter, Saturn and Venus to a few percent accuracy, in astronomical units. Since I know I can do that, there seems no point in not using more precise values from a computer program. Jupiter was at a distance of 4.0686 a.u. when I took the photographs, and its equatorial diameter then turns out to be $(9.67 \pm 0.20) \times 10^{-4}$ a.u. Since I know the size of the Sun in these units, I can calculate that

Physical properties

the equatorial diameter of Jupiter is 0.104 of the Sun's diameter, and its volume, allowing for the oblateness is 1.029×10^{-3} of the Sun. In the next chapter I use an analysis of Jupiter's moons to show that the mass of Jupiter is $(9.55 \pm 0.08) \times 10^{-4}$ of the mass of the Sun, and therefore, dividing the mass by the volume I find that the density of Jupiter is 93% of the density of the Sun. This is fascinating stuff! Jupiter and the Sun are made of much the same materials. I showed in the previous chapter that I could find the density of the Sun to be 1.42 g/cm^3, without assuming a value of the astronomical unit, and it follows that the average density of Jupiter is 1.32 g/cm^3. If I accept the handbook value for the astronomical unit, I can calculate that the equatorial diameter of Jupiter is $(144 \pm 3) \times 10^3$ km, and its mass is $(1.90 \pm 0.02) \times 10^{27}$ kg.

I planned to photograph both Saturn and Mars at their oppositions in 2014. A long streak of really bad weather got in the way, so both photographs were taken sometime after opposition. I did take accurate distances from a planetarium program, but both images were a little smaller than ideal

.

I photographed Saturn on the night of May 18, 2014. I played with the brightness and contrast settings for the image to produce best defined edges. The equatorial diameter of the planet was just 40 ± 3 pixels, and this converts to (19 ± 1.4) arc-seconds. Its distance was 8.9099 a.u. so that its diameter was $(8.2 \pm 0.6) \times 10^{-4}$ a.u., or 0.088 ± 0.006 times the diameter of the Sun. If I use the Handbook value for the astronomical unit, this is $(1.2 \pm 0.1) \times 10^5$ km. I have been a bit cautious with the final uncertainty. This is the diameter of the body of the planet. The rings had an outer diameter of 88 pixels, which translates into 42 arc-seconds, or 2.7×10^5 km.

I had actually photographed Mars a week earlier, on May 11, 2014. This was well after opposition and the planet was beginning to show shading on one side. The image was not quite circular. From measurements on several images with different exposures, I arrived at a polar diameter of 30 ± 4 pixels, and this converts into 14 ± 2 arc-seconds for the angular diameter. The distance was 0.6860 a.u., which leads to a diameter for Mars of $(4.7 \pm 0.7) \times 10^{-5}$ a.u., or 7000 ± 1000 km. This is also 0.0051 ± 0.0008 times the diameter of the Sun. These various versions of the diameters may seem not to convert exactly because I have rounded them down to the significant figures.

For the planets, I had mixed results using the timing method. The difficulty, of course, is that the times are much shorter than for the Sun or Moon, and the random human errors in starting and stopping the stopwatch are correspondingly more important. The way to minimize this is to take a lot of measurements and average the values. The average and its uncertainty are calculated using formulas given in Appendix C. Even using this approach, I found that when the times got shorter than 1 second, the scatter due to this human became the dominant effect. Not only was the scatter relatively large, but it seemed that I was consistently overestimating the short times. Notice that you can't get around this by using a higher magnification. I did, in fact, feel most comfortable using my highest magnification eyepiece, but the image moves faster across the field of view proportionately to the magnification and the time interval is unchanged.

For Jupiter and even Saturn, the final results were comparable with the photographic method. I made a series of timings of Jupiter on September 16, 2009, and found an average time for the planet to disappear from the field of view of (3.3 ± 0.15) seconds. The declination of the planet was -16° 24' so that the time converts into an angular size of (47.5 ± 2) arc-seconds. At its distance of 4.1744 a.u. this corresponds to a diameter of 9.64×10^{-4} a.u. This is very close to the photographic result. There was no possibility with this method of measuring the oblateness of Jupiter.

I made timings of Saturn on several nights in 2010 and 2011. A good set of results was taken on April 26, 2011. The average of thirteen measurements of the time was 1.22 seconds, with an uncertainty of 0.13 seconds. The declination of Saturn was very small, so the cosine correction was negligible. The time then translates into an angular size of (18 ± 2) arc-seconds. Saturn was at a distance of 8.6944 astronomical units, so the angular size corresponds to a diameter of $(114 \pm 12) \times 10^3$ km. This is a diameter for the body of the actual planet, not including the rings.

For the other planets, the timing method gave poor results. I did take a series of measurements of Mars on January 27, 2010. There was a lot of scatter in the times, which were just above 1 second. The average time was 1.3 ± 0.2 seconds, which translated into an angular size of 18 ± 3 arc-seconds. This is definitely larger than the photographic method, and, because of the scatter, I prefer to go with the smaller value. For Venus, the difficulty is that when in it is relatively large in the sky, its shape is half or less illuminated, and it is difficult to make a well-defined timing measurement. I did attempt a series of measurements of Mercury, but the times were under a second, again with a lot of spread, and I didn't have any confidence in them. With the mechanical stopwatch that I was using, the time for me just to start and stop the watch is about 0.5 seconds, so I feel my reaction time is going to be large part of any result once the time gets below 1 second.

Physical properties

Appendix: Timing Method

The second method I used to find the angular sizes was to measure, with a stopwatch, the time needed for the Sun or planet to exit completely from the field of view, that is, the time between the first and last contacts of the image with the edge of the field of view. You might ask why I should spend time on a second method, but I like the idea that two very different methods of measurement can lead to the same result, and, at least for the Sun, the two methods are competitive. There are a few precautions to be taken to make this a competitive approach. The telescope remains stationary during the process, and it is essential to line up the object in the telescope so that it moves radially across the field of view. The first and last contacts of the object with the field stop are then at exactly the same point. Since the azimuth and altitude of that point are fixed, it follows that the declinations of the two sides of the object are equal. The time that you measure with the stopwatch is the difference in right ascension between the two sides of the object. You convert the time difference into an angle using a conversion factor of 360° for a complete day. This works out to 15° per hour, or 15 arc-minutes per time-minute, or 15 arc-seconds per time-second. There are two corrections to this. The Earth actually turns through 360° in a sidereal day, which is a bit less than 24 hours. More accurately therefore, 1 second of clock time corresponds to an angle of 15.0411 arc-seconds. In principle, this should be used for planetary timings, although the correction will probably be much smaller than the uncertainty in the measurement. In the case of the Sun there is a second correction to be made. It takes just over two minutes for the Sun to pass out of the field of view and its right ascension increases by a small amount during this period. On average over the year it increases by 0.0411 arc-seconds per second of clock time. You need to subtract this from the measured interval. You will notice that to this accuracy the two corrections cancel out! Now you have the difference in right ascension for the two sides of the object, and they both have the same declination. There are general formulas for the angle between two points, given their right ascensions and declinations. The formulas are given in the books "Astronomical Algorithms" by Jean Meeus and "Astronomy with your home computer" by Duffett-Smith that I have referred to many times, but for the small angles that we are dealing with there are powerful approximations that reduce the formula to the simple form

Angular diameter = (difference in right ascensions) × cos(δ)

where δ is the declination.

Chapter 22: Moons and Masses of Jupiter and Saturn

Probably every year since 1610, when Galileo discovered them, someone has followed the motion of the four largest moons of Jupiter during an opposition. My own immediate push to do this came from John D. Clark's book, "Measure Solar System Objects and Their Movements for Yourself". I followed his general ideas, but the photographic technique I used and the method of analysis are sufficiently different that it seems worth describing them. In particular, I was able to push the analysis far enough to give me a value for the mass of Jupiter.

I had put off this particular project in part because it clearly requires the use of a telescope. I have lived in the same development for over forty years and during that time the neighborhood trees have grown to limit my view of the sky. For that reason, I use two small refracting telescopes on azimuthal mounts for most of my viewing. I can pick them up and move them from one clear patch of sky to another. They are probably not the ideal choice for this project, but they are what I had to work with. I simply used my DSLR mounted at the prime focus of a 90 mm aperture, 910 mm focal length refractor, with no provision for tracking. I found that if I set the ISO of the camera to its maximum value of 3200, I could get away with exposures of a fraction of a second, usually 1/3 or 1/5 seconds. This gave images that I could measure to find the distances of the moons from Jupiter, in pixels. I have described earlier the calibration process to convert these numbers into arc-seconds. I did take some care over the times of the photographs. Of course, the digital camera records the time at which the photograph is taken, but I found that the clock in my camera lost about one second per day. Rather than trying to make frequent small adjustments to the clock, I left it unchanged and made the corrections to the time in my analysis. As a result, I feel that I have the relative times at which the photographs were taken accurate to the second.

There was a potential problem, that our line of vision does not lie exactly in the plane of the orbits of the moons, so that it is not just a matter of measuring the straight-line distance from each moon to the center of Jupiter. Dr. Clark's photographic technique allowed him to detect the cloud belts on Jupiter and to measure distances parallel to the line of the belts. In my images, Jupiter itself was very overexposed and I was never able to detect any surface features on it. In practice this didn't seem to matter. Generally I could see all four moons and they established the plane of the orbits quite well. Most often, only the outermost moon, Ganymede, was out of the plane, and I could eye-ball the correction sufficiently well. I checked that in part by repeating the measurements on different occasions. Perhaps I was lucky and had taken my photographs at a time when the line of sight from the Earth was close to the plane of their orbits. I judged that I could measure the distances to an accuracy of about two pixels, or three seconds of arc, and this was supported by the quality of the fits to the data that I was able to make. Figure 22.1 shows a typical photograph, taken on November 19, 2011. The 90 mm telescope doesn't resolve the moons as disks; the images are just diffraction blobs. But I was able to pick out their centers accurately.

Figure 22.1 Jupiter and the Galilean moons, November 19, 2011

A pleasant surprise was that I quickly found that I could distinguish the different moons fairly reliably. I noticed on the first photograph of the series that one moon was fainter than the others. That was, in fact, Callisto, but I didn't need to know that. On the third photograph I realized that I could pick out the brightest moon, and that was Ganymede. By the tenth evening, I decided that I could put all four moons in order. I went back to the first photographs and put all four moons in order for the complete series. Based on continuity, I made only one mistake. So from then on I identified all the moons confidently.

From the photographs I could measure the distances of the moons from Jupiter in pixels. One reason for using the 90 mm refractor was that I had calibrated that telescope – camera combination accurately so that I could convert pixels into arc-seconds. The uncertainty of two pixels translates into about three arc-seconds. However, the ultimate goal is to convert the angles into distances and I do that by multiplying by the distance of Jupiter from the Earth. It is convenient to make the observations at a time when the distance is not changing much, and I wanted to be able to use my own measurement of the distance. Both of these requirements indicate making the observations around Jupiter's opposition. I estimated that the distance of Jupiter from the

Earth changes by 2% in about 23 or 24 days, and this seemed a small enough correction that I could make it accurately. So I took my photographs in a forty-six day period around the opposition of 2011. In that period I managed to take 32 sets of photographs spread over 31 nights. That was about the percentage of clear nights that I had expected. I converted the pixel distances into arc-seconds and added a small correction to the early and late dates to compensate for the extra distance.

The results for Callisto are plotted in Figure 22.2. It is very reasonable to try to fit these points with a sine curve. I wrote this in one of two equivalent forms:

$$y = d\sin\left(\frac{2\pi t}{T} + \varphi\right) = d\sin\left\{\frac{2\pi(t - t_0)}{T}\right\}.$$

Here d is the radius of the orbit of the particular moon, t is the running value of the time, and T is the period. The factor of 2π is appropriate if the

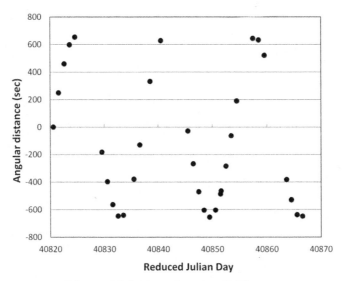

Figure 22.2: Raw data for Callisto

angles are measured in radians. For degrees, it should be replaced by 360°. I need a third parameter which has the effect of shifting the curve backwards and forwards along the time axis. In the first version φ is called the phase angle. One property of the phase angle is that you can add or subtract any multiple of 2π, or 360°, to it without changing the value of y. This is because

Physical Properties

the sine function is periodic; it repeats itself. In the second version I have moved the phase information into a parameter t_0. The significance of t_0 is that it is any value of the time such that the displacement is zero and increasing. I find that easier to think about than the phase angle φ.

From the graph of raw data for Callisto I can read off a value of d as the maximum distance above or below the axis. It is about 650 arc-seconds. There are about 2 ¾ oscillations in the 46 day interval, so the period T is about 16.7 days. For t_0 I can choose any time where the distance is zero and increasing, such as 40820 days. These are a good starting point. The plot for Ganymede was also more or less a sine curve, but the raw plots for Europa and Io were much more difficult to untangle. The raw data for Io is shown in Figure 22.3. I found a neat way to make sense of this picture. If I knew the period of Io, I could reduce all the points to one oscillation, using the "modulo" or MOD function. That is, along the horizontal axis I can plot the times reduced to one period in the form

MOD(time – first time, period).

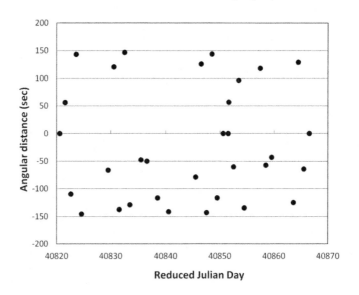

Figure 22.3 Raw data for Io

I have subtracted off the time of the first observation to avoid working with very big numbers. Of course, the period is what I am trying to find, but it is a simple matter in a spreadsheet to assign a guess at the period to some cell, and then keep changing the guess until the curve looks like a sine curve. The plot is very sensitive to the value of the guess. The plot for a trial period of

148

1.768 days is shown in Figure 22.4. The points with zero displacement correspond to the moon being either in front of, or behind, Jupiter. Either ways, I couldn't see it. In the fitting process I gave them uncertainties based on the size of the image of Jupiter in my photographs. Io is somewhere in that range.

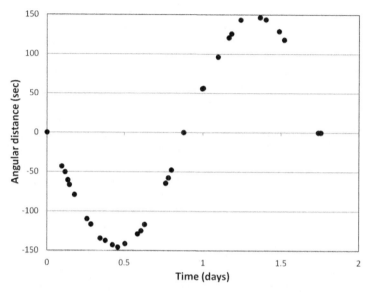

Figure 22.4 Results for Io plotted using the MOD(,) function

I used this way of plotting the date for each of the moons and that enabled me to come up with rather good first estimates of the three parameters for each moon. Once I have an approximate set of values, I can refine them using an iterative non-linear least squares method such as I outline in Appendix D. The results turn out to be very well defined. That is, the uncertainties are very small. The values for the mean distances of the moons from Jupiter are in the table below.

Planet	arc-seconds	a.u.	10^3 km
Io	146	2.808×10^{-3}	420 ± 3
Europa	234	4.498×10^{-3}	673 ± 7
Ganymede	372	7.154×10^{-3}	1070 ± 5
Callisto	656	1.2618×10^{-2}	1888 ± 10

To convert the angular distances into astronomical units I used the distance from Earth to Jupiter in astronomical units that I had measured. To convert these distances into kilometers I had to use the handbook value for the astronomical unit.

149

The periods of the moons, in days, that came out of the fit are in the second column of the next table.

Planet	Measured period	Synodic period	Sidereal Period	Uncertainty
Io	1.7682	1.7701	1.7694	0.0003
Europa	3.5461	3.5541	3.5512	0.0015
Ganymede	7.136	7.1678	7.1560	0.0021
Callisto	16.58	16.75	16.69	0.01

I was initially disappointed with the raw results, given in the second column, in that my numbers differed from the handbook values by many times what I believed to be the uncertainties, shown in the final column. I eventually realized that I was comparing apples and oranges. In connection with our own moon I distinguished between the synodic month, measured relative to the Sun, and the sidereal month, measured relative to the stars. The same two periods can be defined for Jupiter's moons, but what I have measured is neither of them! The periods that are seen from the Earth can be called Earth-based-synodic periods. There is a very good discussion of the various numbers for the four moons in Jean Meeus's book "Mathematical Astronomical Morsels", but I did get led astray for a while by a statement in the book that the mean synodic period seen from the Earth is almost identical with the synodic period seen from the Sun. I am sure this is true, but the key word here is "mean". I eventually realized that the period seen from the Earth varies quite a bit, and the variation is largest just when Jupiter is at opposition. I worked out a simple formula for the correction in this situation. It is given in an appendix at the end of this chapter. The table shows my corrected values for both the true synodic and sidereal periods. The agreement with handbook values is now inside my estimated uncertainties.

The inner three Galilean moons are tied together in something called a Laplacian resonance. This is described nicely in Jean Meeus's book. The three synodic periods should satisfy a relationship

$$\frac{1}{T_{Io}} - \frac{3}{T_{Europa}} + \frac{2}{T_{Ganymede}} = 0.$$

For my three values the sum is -0.00013 ± 0.00038. This is zero comfortably within the uncertainty.

There is a similar relationship between the phase angles

$$\varphi_{Io} - 3\varphi_{Europa} + 2\varphi_{Ganymede} = 180^\circ.$$

The values that came out of the fit are

$$\varphi_{Io} = 49.0^\circ \pm 0.2,$$
$$\varphi_{Europa} = 321.8^\circ \pm 0.7,$$
$$\varphi_{Ganymede} = 188.5^\circ \pm 0.3,$$

and then

$$\varphi_{Io} - 3\varphi_{Europa} + 2\varphi_{Ganymede} = -539.4^\circ \pm 2.2.$$

Remember that the phases are arbitrary within addition or subtraction of any multiple of 360°. If I add 720° to the sum it becomes 180.6° ± 2.2°. This equals the expected value of 180° within the uncertainty.

Finally, it is interesting to look at the four moons as a second example to test Kepler's laws of planetary motion, and this was recognized by supporters of Copernicus's theory as soon as the moons were discovered. When I tested Kepler's third law, in chapter 17, it was convenient to measure the distances of the planets from the Sun in astronomical units, and their periods in years. The ratio a^3/P^2 then comes out as 1. I am going to use the same units to test the law as it applies to Jupiter's moons.

Planet	d(a.u.)	T(yrs)	d^3/T^2
Io	2.808×10^{-3}	4.8443×10^{-3}	$(9.44\pm0.18)\times10^{-4}$
Europa	4.498×10^{-3}	9.7224×10^{-3}	$(9.65\pm0.30)\times10^{-4}$
Ganymede	7.154×10^{-3}	1.9592×10^{-2}	$(9.54\pm0.13)\times10^{-4}$
Callisto	1.2618×10^{-2}	4.5690×10^{-2}	$(9.62\pm0.15)\times10^{-4}$

The four values of d^3/T^2 are equal within their uncertainties. An average value, weighted according to the uncertainties, is $(9.55\pm0.08)\times10^{-4}$. Why did I use these units, and what is the significance of the number that comes out? I have put the algebra in a second appendix at the end of the chapter, but the answer is that this number is the mass of Jupiter measured relative to the mass of the Sun. That is

$$\frac{\text{mass of Jupiter}}{\text{mass of Sun}} = (9.55\pm0.08)\times10^{-4} \approx \frac{1}{1000}.$$

Of course, if I am prepared to use the handbook values for the astronomical unit and the gravitational constant, I can calculate the two masses in kilograms as 1.9885×10^{30} kg for the Sun and $(1.90 \pm 0.02) \times 10^{27}$ kg for Jupiter. In the case of the Sun, my only input was my value for the sidereal year. Aside from the fact that these large numbers are difficult to visualize, it is neat that I could get as far as I did without consulting the handbook. I am rather pleased with myself to have measured the mass of Jupiter with a $300 telescope.

I really enjoyed this project and I recommend it to everyone! An obvious follow-up project is to measure the orbit parameters of one or more of Saturn's moons and use that data to find the mass of Saturn. At the time I was making the observations of Jupiter's moons, I felt that Saturn was outside the range of my equipment. Its brightest moon, Titan, is three magnitudes fainter than Jupiter's moons, and, at magnitude 8.4 is near the limit of what I find I can measure. But in the summer of 2020 I replaced my old Nikon D80 with a D7500 and thought that the greatly increased sensitivity of the camera would make the measurement possible.

It turned out that almost every aspect of the project was changed. Mainly because of very uncertain weather that summer I decided to use the zoom lens set at its maximum of 300 mm instead of the telescope to take the photographs. This did simplify the necessary calibration process but resulted in Titan never being more than 60 pixels from Saturn. Next, both Saturn and even its rings were very over-exposed on the photographs, by at least a factor of 1000. No surface markings were visible and the image of Saturn and its rings was an enlarged slightly fuzzy oval. I found that the most reliable measurements I could make were of the centers of Saturn and Titan. This allowed me to calculate the distance of Titan from the planet but not an accurate track of where it was in its orbit.

Assuming that the orbit of Titan was near enough a circle that I was seeing tilted, I could write down an expression for the distance between Titan and Saturn involving the size, d, of the orbit, the time of the photograph and the period, T, of the orbit, the tilt of the orbit and some measure of the phase, such as one time when Titan seemed to be at its maximum separation. That is four parameters to be fitted. The next question that came up was what was the appropriate period? For the measurements of Jupiter's moons, I could convince myself that the appropriate period was the Earth-centered synodic period, calculated at opposition. I derive a formula for this in the first appendix. I believe this is correct, but the photographs of Jupiter's moons were all taken in a relatively short time span around opposition. In the case of Titan, mainly because of the unfriendly weather, the measurements didn't start until just past opposition and continued for ten weeks. The formula derived in the appendix gives the correction to the true synodic period at opposition, but I know that the correction decreases as the time changes and must be zero at a time somewhere close to three months later, when the Earth is at the quadrature position relative to Saturn (at the top of the circle in Figure 22.5 in the appendix). My solution was to write the formula for the distance of Titan from Saturn in terms of the true synodic period, with a correction for each data point to give my best estimate of the Earth-based synodic period at that time!

This all seemed very complicated and I was pleasantly surprised when a least-squares fit to the measurements converged quickly to a good result. My best results are that the sidereal period of Titan is 15.93 ± 0.02 days, the orbit radius is $(1.20 \pm 0.02) \times 10^6$ km, and the angle of tilt relative to my line of sight is $23.6°$. To find the mass of Saturn I need the sidereal period in years as 0.04362 yrs, and the orbit radius in astronomical units as 8.03×10^{-3} a.u. The mass of Saturn divided by the mass of the Sun is then

$$\frac{\text{mass of Saturn}}{\text{mass of Sun}} = 2.72 \times 10^{-4}$$

Putting in the mass of the Sun, the mass of Saturn is $(5.4 \pm 0.2) \times 10^{26}$ kg.

One reason I wanted to find the mass of Saturn was to verify for myself its overall density. I reported its diameter in the previous chapter as 1.2×10^5 km and the density is then the mass divided by the volume, which gives 6.0×10^{11} kg/km^3 or in more familiar units 0.60 g/cm^3. This is somewhat below the handbook value and, of course, my number has a good sized uncertainty associated with it. But in fact most of the discrepancy is because in the photographs that I took of Saturn in 2014 its rings blocked the north and south parts of the disk so that I couldn't get a value for the polar diameter or the oblateness. In fact, the oblateness is 0.1, so that its polar diameter is only 0.9 of its equatorial diameter. Putting this into the calculation would increase my value for the density to 0.66 g/cm^3. In either case the value is well below the density of water, of 1 g/cm^3, so that Saturn would float in your bath-tub, if you could find a big enough bath-tub!

Physical Properties

Appendix 1: Synodic and sidereal periods

This formula is accurate only close to opposition, but that is what I need.

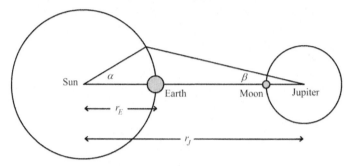

Figure 22.5 Comparing the synodic and Earth-based synodic periods

In the diagram, the moon is any one of Jupiter's moons. r_E and r_J are the distances of the Earth and Jupiter from the Sun, so that the distance of Jupiter from the Earth is $r_J - r_E$. Suppose that I start measuring time from the moment when the moon is exactly in line with the Earth and Jupiter. At a later time t, the Earth has moved through an angle

$$360^o \times \frac{t}{P_E}.$$

P_E is just the Earth's sidereal year. Of course, Jupiter is moving also and it moves through an angle

$$360^o \times \frac{t}{P_J}.$$

I didn't show the movement of Jupiter in the diagram in order to keep it simple. So the picture is drawn in a reference frame that rotates with Jupiter. Because of that, the angle α is actually the difference

$$\alpha = 360^o \left(\frac{t}{P_E} - \frac{t}{P_J} \right).$$

For small values of t, the short arc of its orbit that the Earth has moved through is near enough a straight line, and the angle β is given by

$$\beta = \alpha \times \frac{r_E}{r_J - r_E} = 360^o \times \left(\frac{t}{P_E} - \frac{t}{P_J} \right) \frac{r_E}{r_J - r_E}.$$

In the meantime, the moon is moving counterclockwise around Jupiter and has turned through an angle

$$360^o \times \frac{t}{T_{syn}}.$$

154

The period here is the synodic period rather than the sidereal period because the whole diagram is rotating round the Sun. As the moon rotates, the Earth is moving to meet it, so the moon will next line up with the Earth when

$$360^o \left(\frac{t}{P_E} - \frac{t}{P_J} \right) \frac{r_E}{r_J - r_E} + 360^o \frac{t}{T_{syn}} = 360^o.$$

The solution of this is

$$\frac{1}{T_{syn}} = \frac{1}{t} - \left(\frac{1}{P_E} - \frac{1}{P_J} \right) \frac{r_E}{r_J - r_E}.$$

At the opposition of 2011, Jupiter was 4.961 astronomical units from the Sun. This leads to the values for the synodic periods given in the table. Once you have the synodic periods, the sidereal periods are found from

$$\frac{1}{T_{sid}} = \frac{1}{T_{syn}} + \frac{1}{P_J}.$$

In the case of Saturn's moon, Titan, the observations were spread over a longer time and it was necessary to solve for the angle β exactly.

Physical Properties

Appendix 2: The masses of the planets

I used the basic equation describing the orbit of an object rotating about a central mass M in chapters 17 and 20. I can write it as

$$\frac{r^3}{T^2} = \frac{GM}{4\pi^2}.$$

For the planets moving round the Sun,

$$\frac{a^3}{P^2} = \frac{GM_S}{4\pi^2},$$

and for a planet's moons

$$\frac{d^3}{T^2} = \frac{GM_J}{4\pi^2}.$$

I am reserving the letters a and P for the planets and using d and T for the moons. M_J is the mass of either Jupiter or Saturn. Divide the left side of the last equation by the left side of the previous one and, to preserve the equality divide the right side of the last equation by the right side of the previous one. This gives

$$\frac{d^3/T^2}{a^3/P^2} = \frac{GM_J/4\pi^2}{GM_S/4\pi^2} = \frac{M_J}{M_S}.$$

The factors of G and $4\pi^2$ cancel whatever units are being used. With the particular choice of astronomical units and earth-years as the units of length and time, the ratio a^3/P^2 is just unity, leaving

$$\frac{d^3}{T^2} = \frac{M_J}{M_S}.$$

Part IV. Mainly Optics

Chapter 23: One to get started - Solar Spectrum

This project grew out of a more vaguely defined physics project. The first systematic study of spectra and their production in a prism was made by Isaac Newton. Indeed, he invented the name "spectrum". His book "Opticks" is a very readable account of his experiments with light, much more readable than his better-known book on mechanics "The Mathematical Principles of Natural Philosophy". In the "Opticks", he describes his studies of the spectrum produced from a beam of sunlight. He was as much interested in the properties of the glass prism that was producing the spectrum as he was in the sunlight, but I have always asked myself why he didn't detect the dark absorption lines in the Sun's spectrum. They were first reported by Wollaston and then studied in detail by Fraunhofer. We now call them the Fraunhofer lines. There is a short account of this in the book "Archives of the Universe" by Marcia Bartusiak. My question was whether or not Newton should have been able to see them with his primitive equipment? So I spent some time experimenting with arrangements of slits and prisms to see how easy it was to find the Fraunhofer lines. However, nowadays very good quality inexpensive diffraction gratings are available and the project gradually morphed into using a grating to look for the absorption lines. By the way, I believe that it would have been very difficult for Newton to see the absorption lines with the set-up he was using and, of course, he had no reason to suspect that they might be there.

During my teaching career, I have often seen the absorption lines in the Sun's spectrum, using a commercial spectroscope. I was interested to see if I could get away with much cruder apparatus. The Astronomical Society of the Pacific publishes a handbook, "The Universe at your Fingertips", that describes many experiments for young astronomers. It includes a simple spectroscope consisting of a short cardboard tube with a slit at one end and a diffraction grating at the other. Similar set-ups have probably been described in other places. My thought was simply to make a bigger version. I used a tube that was about three feet long and blackened on the inside. The diffraction grating was cut from a sheet with 1000 lines per millimeter. I did cut the grating-end of the tube at an angle so that I would be looking straight through the grating at the blackened wall of the tube.

The handbook states that the narrower the slit, the sharper the spectrum. For the short tube that they suggest, this is a near-enough correct statement. I had problems initially, because I had assumed that the same idea would apply to my longer tube. Eventually I gave myself a mental kick in the pants. There is an optimum width of the slit for a given tube length. If the slit is too narrow, diffraction becomes important and the spectrum gets fuzzy. Exactly the same idea applies when you are taking photographs through a pinhole. I describe that in the next chapter, but I actually did the thinking about that and the photography several years earlier. Why didn't I catch on sooner? The formula is very slightly different for a slit geometry than for a pinhole, but the result is almost the same. The optimum slit width is essentially the square root of the product of the average wavelength of light and the distance from the slit to the grating. For my tube this worked out to about 0.5 mm. For a much shorter tube, it would be an impractically narrow slit, so the statement in the handbook is correct. I tested the set-up by looking at the spectra of a tungsten light bulb and a compact fluorescent bulb, which are both good spectra to show in an indoor demonstration, and then took the spectroscope outdoors.

Once I had the best slit width, the spectroscope functioned very well visually. Even though I was looking through a narrow slit, I preferred not to point the tube directly at the Sun. I set up a white screen in sunlight, and pointed the tube at that. I threw a black cloth over my head and the business end of the spectroscope, because the spectrum was quite dim due to the narrow slit. I could easily see a large number of dark lines in the spectrum. We always refer to "lines", but they are only lines because the slit is straight. If the slit was curved so would be its images formed by the different colors.

I had more of a struggle than I had anticipated taking photographs of the spectrum that clearly showed the lines. At least a part of the difficulty arose from the different responses of our eyes and the sensor in a digital camera to the different colors. The three types of receptors in our eyes do have peak sensitivities in the red, green and blue regions of the spectrum, but they have rather large pass-bands. The filters in a digital camera have much sharper cut-offs. That is certainly the case for my Nikon D80. One region this shows up is in the yellow part of the spectrum. The light reflected from yellow paint is a mixture of red and green colors, and the algorithm used in the camera to form the yellow image from the responses of the red and green pixels does a good job. If a single pure wavelength of light in the yellow region hits the sensors, they don't cope with it so well. At these wavelengths the sensitivities of both the red and green receptors are dropping off precipitously. Depending on the ISO setting and the exposure, I sometimes recorded a very dark band and at best the yellow region of the spectrum came out narrower than our eyes see it. A second difference comes at the short wavelength, or blue, end of the spectrum. Isaac Newton famously noted indigo and violet colors after the blue part of the spectrum. That depends on the red sensors in our eyes having a long tail of sensitivity that stretches right through the blue region. The sharp cut-off filters in the digital camera don't do that and the short wavelength end of the spectrum is simply blue.

I photographed the spectra by mounting my camera, with the 50 mm lens, on a separate tripod, pointing into the diffraction grating in about the right direction, and throwing the black cloth over it to keep out stray light. I tried a variety of ISO settings and exposures, as well as other focal length lenses, and Figure 23.1 shows the best result I have made so far. On the original photograph there are many lines visible. The black and white version has been slightly sharpened. I guessed that the prominent line at the left side was the α line of hydrogen, called the C line by Fraunhofer, that the line at the beginning of the orange region was the D line of sodium, and that the dark line at the right end was Fraunhofer's G line, an overlapping of iron and calcium lines. Remember my comment that the digital camera has problems with yellow light. In the color version of this photograph, the sodium line is distinctly more orange than it appears to my eyes, and there is a rather dark region to the right of it, where the sensor can't cope with the yellow wavelength. From the positions of the C, D, and G lines I could see that the relationship between the wavelength and the number of pixels from the left end was very nearly linear. In an effort to get the best accuracy from the results, I used Excel to fit a quadratic expression to the three lines, but it made only a small difference. I could then calculate the wavelengths of the other lines, and I could identify a number of other lines as listed in tables,

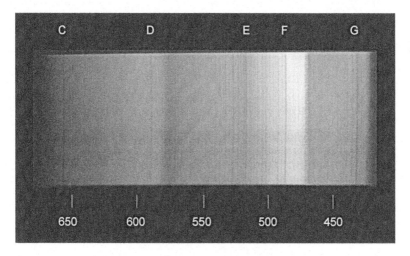

Figure 23.1 Solar spectrum

such as you can find in Wikipedia. The scale in nanometers is shown on the photograph. I could find the hydrogen lines including the Hβ labelled F, and the Hγ just to the left of the G line, at 434 nm, and several lines from iron and magnesium. A difficulty is that my resolution is a few nanometers, and there are just a lot of lines that can be this close to each other. However, I had achieved as much as I set out to do, so I am satisfied with the results.

I can almost hear your question "What about the spectra of stars?" I have only one preliminary result to show, of the spectrum of the brightest star, Sirius. I simply removed the slit from the front of the tube spectroscope, and relied on the point-source appearance of the star to give the resolution in the spectrum. The tube was there to provide some shielding from other light sources. I waited until Sirius was crossing my meridian, so that it was moving horizontally through the sky, and oriented the tube and grating so that the spectrum was vertical. That way, the movement of Sirius during a long exposure has the helpful effect of broadening the spectrum into a band, without smearing out any detail. The resulting photograph, Figure 23.2, is a full-color spectrum, but it doesn't show any dark lines. There is a web page www.eso.org/~rfosbury/home/natural_colour/spectrophot/nc_spec_slitless. html (in 2020) which shows a spectrum of Sirius taken with a similar arrangement to mine, and which does show absorption lines. I note that Dr. Fosbury used a medium format camera, with a correspondingly larger-diameter lens, and it may be that this gave him better resolution. The strongest claim I can make for my result is that the light that formed this spectrum did come from the star Sirius!

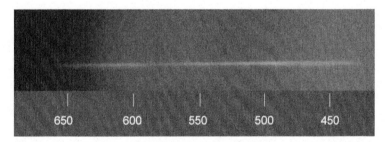

Figure 23.2 Spectrum of Sirius

The spectrum does change from a red band, through a narrow yellow band, and a green band to a blue band, though you will have to take my word for that. I have superimposed the nanometer scale from the solar spectrum since the scales of the two photographs are the same.

Chapter 24: Pinhole images of Sunspots

Sunspots were studied intensively almost immediately after the development of the telescope. Certainly Galileo looked at them, though there is some controversy if he was the first person to do so. A question is whether sunspots had been observed without a telescope. The idea of this project is to use a lensless pinhole optical system to image sunspots. The technology to do this has existed for thousands of years. I first read of this possibility in a long-lost astronomy book several decades ago, but didn't get around to trying it until 2005. I did find an earlier very similar project to mine on the internet at users.erols.com/njastro/barry/pages/pinhole.htm. There is also an account of the same method used to view the 2004 transit of Venus at www2.eng.cam.ac.uk/~hemh/transit.htm.

The idea appealed to me because I have always been fascinated by the concept of a pinhole camera. In elementary school I wrote a book report on one and in high school I actually built one and took at least one photograph with it, for a science project. The basic idea is simple. Light strikes each point in an object, such as the arrow on the right of the picture. Reflected rays of light spread out from each point in straight lines. A small opening in the side of a box selects a small bundle of rays which strike the other side of the box forming a slightly fuzzy image.

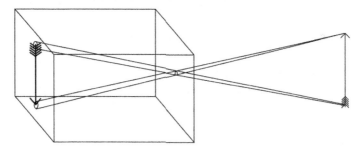

Figure 24.1 Principle of the pinhole camera

It sounds crude, but the results can be surprisingly good. Figure 24.2 is a pinhole photograph I took in my backyard some years ago. Part of the trick is to make the camera as large as possible. This photograph used a piece of 8 × 10 inch enlarging paper as the sensitive emulsion. Digital cameras are at a disadvantage because of the small size of the sensors. I have made pinhole images using a DSLR with the lens replaced with a pinhole, but the resolution could not match the large-camera images. The second part of the trick to making sharp photographs is to use a pinhole whose size is

Figure 24.2 Pinhole photograph

optimum for the camera dimensions - more on that below. The pinhole camera has had enthusiasts over the years. Two books on the subject are "The Hole Thing" by Jim Shull, and "Pinhole Photography – Rediscovering a Historic Technique" by Eric Renner, both listed in the references.

Here are three astronomical examples of pinhole images. In 1970, I was living in Ontario and there happened to be a partial eclipse of the Sun. I didn't own a telescope at the time so the evening before the eclipse I put together a very simple pinhole device. It was just a cardboard tube with a piece of aluminum foil taped across one end, with a pinhole in it, and a piece of tracing paper at the other end. I cut a cardboard box to fit on the lower end to provide some shade for the image. Figure 24.3 shows the set-up and Figure 24.4 shows the view looking up into the box. It worked, but if I was in a similar situation now I would use the type of arrangement that I describe below for looking at sunspots.

The second astronomical example was pure serendipity. I stepped onto my front porch one day and saw the pattern of light shown in Figure 24.5 on the wall. Each circle of light is an image of the Sun, formed by rays of sunlight passing through a natural "pinhole" formed by a chink between the leaves of a tree. This example does give a hint of a problem to come.

Figure 24.3 Pinhole set-up to photograph eclipse

Figure 24.4 View of the partial eclipse

Mainly Optics

The tree was about thirty feet away and the circles of light are about three inches in diameter. But notice that the images are round, independent of the shapes of the chinks.

Figure 24.5 Natural pinholes

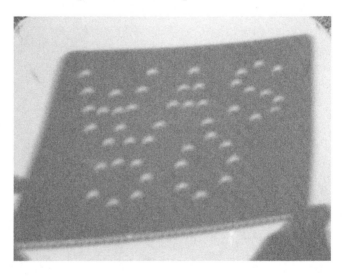

Figure 24.6 WAS Anniversary!

For the third example, in August 2017 I travelled to Glendo, Wyoming to watch the total eclipse of the Sun. Before going I had prepared a sheet of card with 47 circular holes punched in it to celebrate the fact that my astronomy club, the West Jersey Astronomical Society, or WAS, was fifty years old. During the partial phase of the eclipse I held up the card to cast a shadow.

I have put the derivation of the formula for the optimum pinhole size in an appendix at the end of the chapter, but the key results are that the best size for the pinhole is related to the distance D of the pinhole from the screen and the wavelength λ of the light by

$$d = \sqrt{2.44\lambda D},$$

and the resolution is given by any of three expressions

$$\frac{d}{\sqrt{2D}} = \sqrt{\frac{1.22\lambda}{D}} = \frac{1.725\lambda}{d}.$$

Provided d has the optimum value, these are all equivalent.

To get some numbers, let's start with the resolution. The Sun is about half a degree across in the sky, and sunspots are often one hundredth of that. If the resolution is 0.005 degrees, which is about 9 x 10^{-5} radians, and I take the wavelength as a middle sort of value of 500 nanometers, or 5×10^{-7} m, I get good news and bad news. The good news is that the diameter d of the pinhole is about 1 cm across - not really a pinhole at all. The bad news is that the distance D from the pinhole to the screen has to be 75 meters, or about 250 feet!

Let's try again, a bit more modestly. For reasons that will emerge, I start by choosing D to be 100 feet, or 30 meters. Then the size of the pinhole is reduced to 6 mm. There is some loss of resolution, but it is still about 1/60 of a degree, which is not bad. There is still the question of how you make a pinhole camera 30 meters long. The answer, like all good conjuring tricks, is that you do it with mirrors.

Here is a first experiment. In Figure 24.7, I am holding an old mirror and using it to reflect a spot of sunlight on to the wall of my house. Figure 24.8 shows what I saw on the wall. I like this pair of photographs because the mirror happened to be square whereas the spot of light on the wall is round, although admittedly quite fuzzy. The mirror is acting as a pinhole, selecting a bundle of rays from the Sun to form an image on the wall.

Figure 24.7: Me holding a square mirror

Figure 24.8 A fuzzy, but circular, patch of reflected light

Things rapidly got more elaborate. Figure 24.9 shows a front surface mirror, intended for use as a diagonal in a Newtonian telescope, mounted on a ball and socket head that is itself mounted on my old AZ3 tripod, which has its own adjustments. The idea is to use the ball and socket head to get the image of the Sun more or less where I want it, and then to use the slow motion controls on the AZ3 to keep the image in one place as the Earth turns. The mirror is, of course, much larger than the 6 mm diameter that I calculated, but it forms a bright reflected image. What I did was line up the

Figure 24.9 The plane mirror on its mount

image using the whole aperture of the mirror and then swing into place the mask shown in the photograph in Figure 24.10, which has a 6 mm hole in it.

Figure 24.10 The 6 mm mask swung into place

Figure 24.11 The complete arrangement

In quite a real sense, I used my house as the camera. Figure 24.11 shows the complete arrangements outside of the house. I have set up the mirror across the street from my house so that the reflected sunlight goes in at the front door, passes through the center of the house and hits a screen towards the back. The mirror then turns out to be just 100 feet from the screen, which is the reason I chose that number. On the original photograph you can clearly see the circular image of the Sun through the doorway. This photograph was taken without the mask in place. With the mask, the image is much fainter and can't be seen from across the street. Clearly, two people are needed; one observes the image on the screen and calls out directions to the person across the street to keep the image centered. You also can't see in the photograph the quite elaborate hangings of black material that were needed to darken the interior of the house sufficiently!

All that remained was to wait for some reasonable sunspots to appear. I waited through the summer of 2005 and on September 10th a good sized group appeared. I looked at them through a white light solar filter on my refractor and took the photograph on the left below. The image on the right is the pinhole result, photographed off the screen. The image of the Sun's disk was about eleven inches in diameter.

Figure 24.12 Sunspots photographed through a telescope and with a pinhole

So it can be done. I am still intrigued by the idea that someone may have done this hundreds or even thousands of years ago. Certainly in the middle ages there were people such as Roger Bacon who were quite skilled in optics and could have tried it. The book "Galileo's Glassworks" by Eileen Reeves contains a comprehensive history of the telescope and related devices up to the time of Galileo, but it has no record of the pinhole approach.

Appendix: Optimizing the size of the Pinhole

Based on the idea that the pinhole simply selects a narrow bundle of rays of light, you might expect that, the smaller the pinhole is made, the sharper the image will be. In fact, if the light is coming from a distant source, so that the light rays are almost parallel, the geometrical image of a point object would just be equal in size to the diameter of the pinhole:

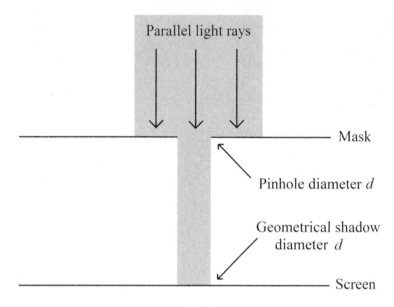

Figure 24.13 The geometrical shadow

However, there is a second effect that gets in the way. When light passes through a small opening it spreads out through the process called diffraction. Actually, most astronomers are aware of this because it puts a limit on the resolution of a telescope. Figure 24.14 shows two photographs of (red) laser light that has gone through a small pinhole to a screen several feet away. The left hand image is a short exposure, showing that the narrow laser beam has broadened into a spot, and there is a faint ring outside it. The right hand image is a longer exposure. In fact the spot and the first diffraction ring are overexposed and have merged together, and now you can see that there is a series of even fainter rings outside. In astronomical situations it is usually only the first ring that can be seen, and in calculating the resolution I am concerned only with the size of the central spot.

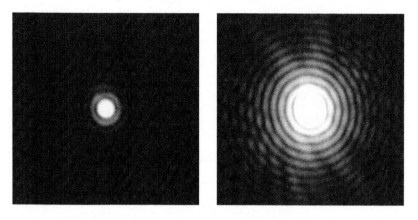

Figure 24.14 Diffraction rings from a laser source

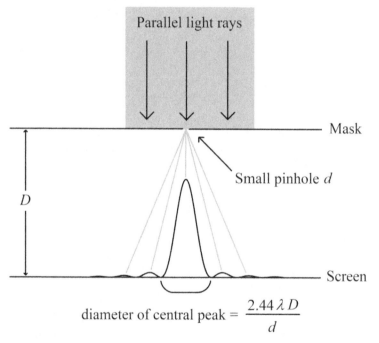

$$\text{diameter of central peak} = \frac{2.44\,\lambda\,D}{d}$$

Figure 24.15 Diffraction broadening of the pinhole image

The diameter of the first zero in the intensity of the diffraction pattern involves the distance D from the pinhole to the screen, the wavelength λ; of the light, and one over the diameter of the pinhole. There is also a mysterious number 2.44. A half of this, or 1.22, is encountered in the Rayleigh criterion for the resolution of a telescope. It comes out of the mathematical summation of all the contributions to the image.

171

A key result here is that the geometrical width of the spot is just d, so that it gets larger as d increase, whereas the diffraction pattern width goes as $1/d$, so that it is largest when d is small. Adding the two contributions together in a rigorous way is rather difficult, but you can make a good guess at the answer.

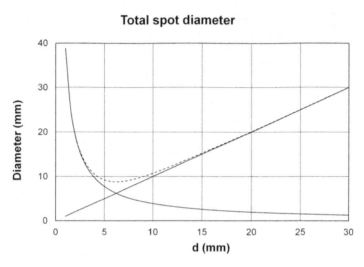

Figure 24.16 The optimum pinhole diameter

I have plotted the two contributions, for a particular choice of D and λ, as the two solid lines. The dashed line is my guess as to the sum. It is actually the square root of the sum of their squares, but simply adding them together would give a very similar result. You can see that there is a minimum in the dashed curve, which represents the optimum value, and that the minimum comes at a value of d close to where the two solid lines cross, or where the two expressions have equal values. This lets me solve for the optimum pinhole size.

$$d = \frac{2.44\lambda D}{d}$$

$$\therefore d^2 = 2.44\lambda D$$

$$\therefore d = \sqrt{2.44\lambda D}.$$

The resolution must also have contributions from both effects. Equivalent expressions that must be about right are:

$$\frac{d}{\sqrt{2D}} = \sqrt{\frac{1.22\lambda}{D}} = \frac{1.725\lambda}{d}.$$

These expressions are all equal when d has the optimum value. Again, this is an approximate value. I am not worrying about a factor of two or so.

Chapter 25: Building and using a Galilean telescope

In 1609, word reached Galileo that a new optical device was being shown in northern Europe, one that showed a magnified image of distant objects. He quickly figured out how to make such a device and created the instrument that we now call a Galilean telescope. He later claimed that he had, in a certain sense, invented the telescope, since he had figured it out by reasoning whereas the previous telescopes had depended on accidental discovery. But Galileo's knowledge of optics was primitive and the reasoning that he gave was at best incomplete. The telescopes that he built utilized a negative lens as the eyepiece. Modern astronomical telescopes use a positive eyepiece design that is named after Kepler, who proposed it in 1610. Kepler seems to have had a better grasp of geometrical optics than did Galileo. It would be interesting to know if the telescopes that had been demonstrated earlier, in Holland among other places, were of the Galilean or Keplerian design. In his book "The History of the Telescope", Henry C. King says that we simply don't know the answer. The information has been lost. It seems likely to me that they were Galilean, because of the property that a simple two-lens Keplerian telescope gives an inverted image. For terrestrial use this would have been a fatal disadvantage. A fascinating modern account of Galileo's life is "Galileo's Daughter" by Dava Sobel.

2009 was, of course, the 400th anniversary of these events, and it was named the Year of Astronomy. I would like to claim that I built my telescope to mark that occasion, but I actually began to work on it in 2007 and it took me a year and a half to get all the results. Actually, it is quite straightforward nowadays to build an optical replica of Galileo's telescope. There are many sites on the internet that offer advice and detailed instructions. Unfortunately, web pages come and go. The two web pages that I referred to in the first draft of this book have disappeared. The April 2009 issue of "Astronomy" magazine also contains some advice. I do recommend that you get to look through such a telescope. For anyone used to a modern refractor, the first view is an eye-opener in more than one sense. How on Earth did Galileo manage to see as much as he did?

The reason for the surprise and the main points of the geometry of the Galilean telescope are shown in Figure 25.1. The eye and the negative eyepiece are to the left. The positive objective is on the right. Whether you like it or not, the eyepiece forms an image of the objective lens and this is shown as the oval shape between the two lenses. The significance of this image is that it serves as the exit pupil for the telescope, which means that every ray of light leaving the eyepiece appears to come from it. A Keplerian telescope has a similar behavior but with the difference that the exit pupil of the Keplerian telescope is outside the eyepiece so that the eye can be placed next to it. This is usually the best place to put your eye. In the case of the Galilean telescope, the eye must necessarily be to the left of the eyepiece and some distance from the exit pupil. This leads to a tunnel vision effect which is very severe for a telescope with the specifications of those Galileo actually used for his discoveries.

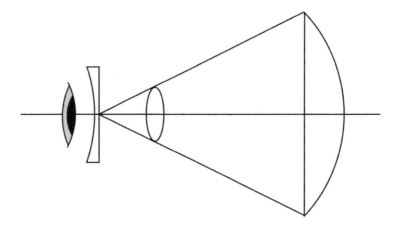

Figure 25.1 Schematic of Galilean telescope

I want to make two points that have not been sufficiently emphasized. The first point is that there are two important limiting regimes for the design. In a low-magnification telescope the exit pupil is typically much larger than the opening of the iris of the eye, and the field of view is then proportional to the size of the exit pupil and hence to the diameter of the objective lens. However, at high magnifications the exit pupil is often much smaller than the iris of the eye and the field of view is then proportional to the diameter of the iris of the eye. I have put in an appendix a fairly mathematical derivation of the expressions for the field of view in the different limits, and I have taken a set of photographs illustrating the idea of a field stop in the two limits. Galileo's first telescope, and the small telescopes being sold in northern Europe, belonged to the first category but, as Galileo increased the magnification of his telescopes, they moved into the second category. If Galileo had experimented with his first telescopes he would have found that the field of view was proportional to the diameter of the objective. He clearly believed this when he wrote his first book, "The Starry Messenger". He would not have realized immediately that it was not the case for his later telescopes. The second point is that, in either limit, you can see a wider field of view by moving your eye around behind the eyepiece without resetting the telescope so that the accessible field of view is not very different from that of a Keplerian telescope. Much of this information is available in different places but I found it helpful to derive all of the results myself using a single notation. The mathematical derivations are in an appendix to the chapter.

The results can be brought together using the concept of a field stop. Every optical system has in it some opening that determines the field of view, and that is called the field stop. For a Galilean telescope, there are three limits of behavior corresponding to different apertures in the system acting as the field stop. In the low magnification limit the field-stop of the system is the objective lens whereas in the high magnification limit the field-stop is the pupil of the eye. If the eye is allowed to move around, or if the eyepiece is small, the field stop is the edge of the eyepiece. I have taken two series of photographs that illustrate the different behaviors. To do that, I needed to make two very different telescopes. The first is a low magnification telescope shown in Figure 25.2.

Figure 25.2 Low magnification Galilean telescope

The telescope is constructed very simply from a cardboard mailing tube, lined with black flock paper. The objective has a focal length of 500 mm and the eyepiece is a negative lens with a focal length of -170 mm. The magnification is therefore approximately 3. Both lenses have diameters of 50 mm. This is almost certainly larger than was available in good quality in 1609, but I wanted to make sure that the exit pupil was large. In fact, the exit pupil is 17 mm in diameter, much larger than either a daytime adapted eye, about 3 mm, or a night-adapted eye, about 6 mm.

Figure 25.3 shows a series of photographs that I took through this telescope, in daylight, simply by putting a camera on a separate tripod behind it. The camera had a 50 mm lens mounted on it, with the aperture set to f16. The diameter of its iris opening is then 50 mm / 16, or approximately 3 mm. This is much less than the size of the exit pupil and corresponds to a daytime adjusted human eye. The left-hand photograph is taken through this set-up. I included in the view an object, a circular planisphere. The planisphere is 16 inches across and it was 160 feet from the telescope, so it has an angular size of 0.5 degrees, which is about the same as the angular size of the Moon. This gives a scale to the photograph. I claim that the objective lens is the field stop in this limit. This means that if I change the size of the objective the field of view scales in proportion. In the center photograph, I have put a mask in front of the objective, reducing its diameter by a factor of two, and the field of view has gone down in proportion. If you put a mask over the objective of a modern telescope, the field of view is pretty much unchanged; the image is just dimmer. Just for fun, in the right-hand photograph, I put a star shaped mask in front of the objective and the view is masked the same way. Try this with your modern refractor!

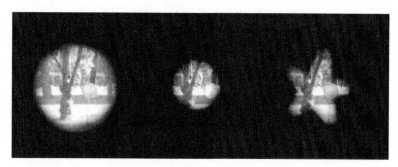

Figure 25.3 Low magnification views with masks over objective

My second telescope, shown in Figure 25.4, is a much more conventional Galilean replica, as far as the optics is concerned. The tube is PVC tubing and the push-pull focuser is cut from a plastic plumbing fixture. I found that lining the tube with black flock paper eliminated internal reflections and made it unnecessary to put in internal baffles. The objective has a focal length of 1000 mm. Its diameter is 25.4 mm, but by the time it was mounted this was masked to about 23 mm. I used two eyepieces, with focal lengths of -50 mm and -30 mm, giving magnifications of 20 and 33 respectively. The eyepiece diameters were also 25.4 mm. The exit pupils with the two eyepieces were 1.15 mm and 0.7 mm, which are both much smaller than a dark adapted eye. The interesting point is that, in this limit, it is the diameter of the eye that determines the field of view.

Figure 25.4 Galilean replica telescope

I took a second series of photographs using this telescope, shown in Figure 25.5. The planisphere still subtends an angle of 0.5 degrees. I used the 20x magnification eyepiece so that the exit pupil of the telescope is 1.15 mm in diameter. This time, the iris of the camera lens is the field stop, and I took the photographs using different apertures for the 50 mm lens on the camera. The telescope was not adjusted at all during the series. For the first photograph, on the left, the aperture is f16 so that the iris opening is about 3 mm, similar to a daytime human eye. The field of view is very narrow. In the second photograph the aperture is f8. The iris opening is about 6 mm, similar to a night adjusted human eye. The field of view is doubled. In the

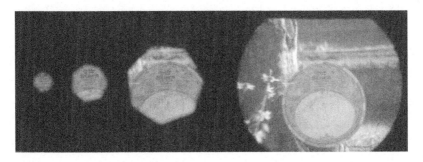

Figure 25.5 Varying iris opening with high magnification telescope

next photograph, the aperture is has been increased to f4, corresponding to a diameter of about 12 mm. Notice that the outer edge of the field of view corresponds to the iris diaphragm of the lens. Finally the lens is wide open at f1.4. The diameter of the iris is about 36 mm. If you could find an alien with eyes this big she would have a fine view!

A more serious comment is that the last photograph gives you some idea of what you can see by moving your eye around behind the eyepiece. Clearly the image of the whole planisphere, and hence of the whole Moon at night, is getting out of the telescope. A related point is that the telescope can also be used to project an image of the Sun, and the image of the entire disk of the Sun is projected.

In this series of photographs I didn't reach the final limit, where the eyepiece of the telescope acts as the field stop. If the camera lens could have been opened much more, or if I had moved the camera slightly to one side, I would have seen the slightly rough edge of the mounting of the eyepiece limiting the view. In fact, based on the limiting formulas in the appendix, I should have reached this third limit when the lens was wide open. The reason I didn't is that the iris diaphragm of the multi-element camera lens is buried deep inside the optics so that I couldn't get it close enough to the eyepiece of the telescope. Perhaps I was being over-protective of the Nikkor lens!

What I Saw

So what did I manage to see through the telescope? First let me emphasize two things that I didn't do, and why I didn't do them. I did not put a finder scope on the telescope and I didn't try to take any photographs of the night sky through it. The reason for both was that I was trying to recreate for myself the experience that Galileo had, as far as is possible in the 21st century. He did not have a finder scope and he certainly could not take photographs. He did make many drawings and I quickly decided that he was a better artist than I am. But that's how it goes. I tried to look at the things that he writes about, to appreciate what they would have looked like to him.

The moon was high on my list of things to look at. Two experiences I had illustrate the differences between Keplerian and Galilean telescopes. When I looked at a bright object like the Moon, I could see something sticking into the field of view at one edge. I took the telescope apart twice trying to find the obstacle. Finally I appreciated what I have written in the theory section. The field stop of the system is the iris of my eye and what I was seeing was an unevenness in the edge of my iris. The second surprise came when I moved my eye around, behind the telescope, to look at the whole disk of the Moon. I could indeed see the whole size of the Moon without resetting the telescope, but there was a rippling of the image as I moved my eye. I assumed that this was an indication of the poor quality of the simple lens I was using as the objective, and I bought a second 1000 mm lens from a different supplier. It turned out to give the same effect. Then I had my wife look through the telescope and she reported no ripple effect.

Again, my ancient eye was to blame. The small exit pupil of the telescope meant that only a narrow pencil of light was entering my eye, much narrower than the size of my iris, and, as I moved my head, the pencil of light passed through different parts of the cornea of my eye. The rippling came from unevenness in my eye. There was nothing I could do about either problem!

The first two drawings were sketched at the telescope. The view was in fact very good, full of detail, much more than I could draw. In the second drawing I did attempt to sketch a small part of the terminator more carefully.

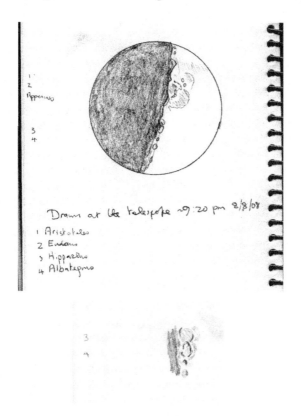

Figure 25.6 Drawings of the Moon made at the telescope

My next target was Jupiter and its moons.

Figure 25.7 Jupiter and its moons drawn at the telescope

Here are tracings from my log book of drawings made over a series of evenings. After completing them, I used "The Sky" to identify which moons were which. I have labeled them with just the first letter of their names: I for Io, E for Europa, G for Ganymede, and C for Calisto. The one surprise was that I thought I had seen four moons on 7/31/08 but one of the dots turned out to be a star.

Venus gave me another share of adventures. Here are notes from my log:

9/24/07: Looked for Venus before sunrise but the eyepiece fell off and I couldn't find it until later!
9/26/07: Did see Venus in a.m. Shape definitely not a circle.
9/29/07: Better look. Venus looks like a banana.

Figure 25.8 "Venus looks like a banana"

10/12/08: Star watch at St. Peter Celestine school. Venus was not a circle. Question from Steve Mattan: Would we have said that if we hadn't known it?

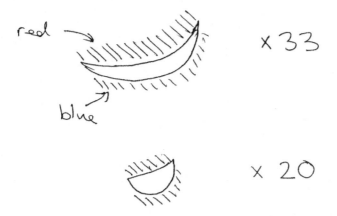

Figure 25.9 Venus sketched at the telescope 2/9/2009

I did finally get to see Venus under better conditions and had some glorious views. My drawings are in Figure 25.9. There was a lot of false color with such a bright subject, but no doubts about its crescent shape at that time.

Mainly Optics

From my log book:
2/9/09: Lot of false color but a very sharp water melon slice in the center of it.
(At this point, I remembered one of Galileo's cryptic messages: "Cynthiae figuras aemulatur mater amorum – Venus imitates the phases of the Moon.")

I was very interested to look at Saturn. It is well known that Galileo did not recognize the ring system but instead thought that there were two companion stars or moons moving with the planet. When the ring system was edge on as seen from the Earth, the companions disappeared. I looked at Saturn on several occasions, using both the 50 mm (x20) and 30 mm (x33) eyepieces. The picture is made up of tracings from my log book.

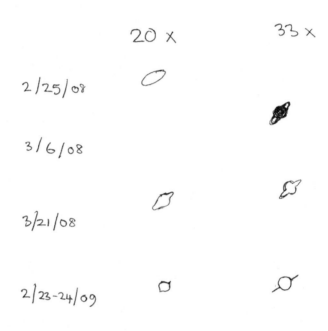

Figure 25.10 Saturn at 20x and 33x magnifications

It seemed to me that the view through the 50 mm eyepiece was more or less consistent with Galileo's interpretation, but that the rings were much better defined with the 30 mm eyepiece. I convinced myself on 3/6/2008 that I could see a space between the upper and lower parts of the ring. Of course, a big danger is that I knew what I was looking at. There is some question of how good Galileo's 30x telescope was and if he actually used it for any useful observations. My experience suggests that it wasn't as good as my telescope used at 33x magnification.

One set of observations that Galileo doesn't always get credit for was of the famous Mizar - Alcor system. Naked eye, or in binoculars, Mizar and Alcor form a double star in the handle of the Big Dipper. In a telescope, Mizar is itself a much closer double star. There is a letter from a former student of Galileo's urging him to look at Mizar. I did observe it with the 30 mm eyepiece and was able to resolve Mizar A and B clearly. I attempted a sketch but the difficulty was that I did have to move my eye to bring Alcor into view and I completely underestimated the distance between it and the two components of Mizar.

Figure 25.11 Alcor and Mizar sketched at the telescope

A web page describing Galileo's observations of Mizar is at http://leo.astronomy.cz/mizar/article.htm. It was also discussed in Sky and Telescope, July 2004, pp 72 - 75.

I also constructed a small screen to hang on the back of the telescope in order to project the Sun's image on it to look at sunspots. The Galilean design works very well in this application. There is certainly no problem with a restricted field of view; the whole disk of the Sun is visible. I assume that Galileo used this method to view the Sun's image safely. The main problem I did have is that I finished making the screen during a very long sunspot minimum so I had to wait many months for a good opportunity to test it. Finally a medium size sunspot group did appear. On the screen they were very easily visible. The photograph is distorted because the camera was looking in from the side. I needed the shade in front of the screen to keep stray light from degrading the image.

Figure 25.12 Projected image of the Sun

Figure 25.13 The Galilean telescope aimed high

Appendix: Derivation of formulas for the field of view

A standard reference for the field of view of a Galilean telescope is a short paper published in 1920, in the Transactions of the Optical Society, by H.A. Hughes and P.F. Everitt. (H. A. Hughes and P. F. Everitt, "On the Field of View of a Galilean Telescope," Trans. Opt. Soc. 22, 15-19 (1920)). In their notation the focal length of the objective is f_1 and the focal length of the eyepiece is $-f_2$. That is, f_2 is a positive number. The magnification m is also a positive number equal to f_1/f_2. The diameter of the objective is $2P$ and the diameter of the iris of the eye is $2p$. The height of the exit pupil is the size of the objective divided by the magnification, or $2P/m$. Their results can be understood in terms of the next, very simplified, diagram, which is a skeletal version of Figure 25.1.

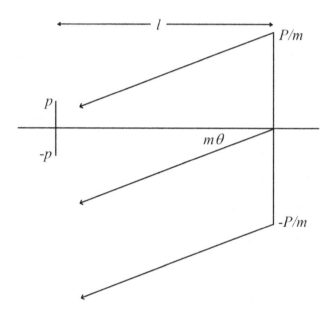

Figure 25.14 Schematic showing field of view of low-magnification telescope

The vertical line on the right represents the exit pupil of the telescope, with a total height of $2P/m$, where $2P$ is the diameter of the objective. The vertical line on the left is the pupil of the eye, with a total diameter of $2p$. The distance between the two lines is labeled l. This must be slightly greater than the distance of the exit pupil from the eyepiece. Later I shall approximate it by this value. Then

$$\ell \approx f_2(1-1/m).$$

As light enters the objective traveling at an angle θ to the axis, rays leave all points on the exit pupil traveling at an angle $m\theta$ to the axis. If the ray leaving the top of the exit pupil passes through the bottom point of the eye pupil, it will be the only ray getting into the eye. This corresponds to an angle given by

$$m\theta \leq \tan^{-1}\left(\frac{P/m+p}{\ell}\right).$$

If θ is reduced until the ray from the top of the exit pupil hits the top of the eye pupil, then the eye will be fully illuminated. The value of θ is given by

$$m\theta \leq \tan^{-1}\left(\frac{P/m-p}{\ell}\right).$$

These are the two extreme cases. Hughes and Everitt also define a median ray corresponding to the ray from the top of the exit pupil passing through the center of the eye. The value is given by

$$m\theta \leq \tan^{-1}\left(\frac{P}{m\ell}\right).$$

These are the results given by Hughes and Everitt. If you completely neglect p compared with P/m, the three formulas become equivalent. If you also make a small angle approximation, $\tan^{-1}(x) \approx x$ in radians, and further replace l with the distance from the eyepiece to the exit pupil, the half-field of view becomes

$$\theta \leq \frac{P}{f_2 m(m-1)} = \frac{P}{f_1(m-1)}.$$

The field of view is directly proportional to the diameter of the objective. This is something that Galileo clearly believed, at least at the start of his investigations.

Hughes and Everitt made it clear that they were interested in field glasses of the Galilean design that were popular before modern prismatic binoculars were introduced. In the examples that they looked at, the exit pupil was larger than the eye pupil. I have a small opera glass made by

Vivitar, described as 4 x 30. The 30 is the diameter of the objective in millimeters. The magnification is 4. The diameter of the exit pupil is then 7.5 mm. In daylight a typical human eye has an opening of about 3 mm so that it is small compared with the exit pupil. The formulas given above are then accurate. This was also very probably the case for the first telescope that Galileo made, with a magnification of 3. However, he rapidly developed telescopes with magnifications of 20 and perhaps 30. The objective lenses were of the order of 25 mm in diameter and the exit pupil was around 1 mm. This situation is qualitatively shown in the next figure.

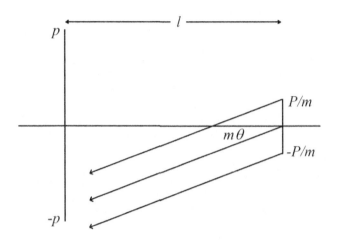

Figure 25.15 Schematic showing field of view of high-magnification telescope

The light leaving the telescope now emerges as a narrow pencil and it is obvious that a dark-adapted eye, with a diameter of about 6 mm, can never be fully illuminated. The condition for any ray at all to enter the eye is unchanged and given in the second equation above. The ray from the top of the exit pupil passes through the bottom of the eye pupil, just as in the previous case. The condition that all the rays leaving the telescope enter the eye corresponds to the ray from the bottom of the exit pupil entering the bottom of the eye. The condition on θ is

$$m\theta \leq \tan^{-1}\left(\frac{p - P/m}{\ell}\right).$$

As a median value, the condition for the ray from the center of the exit pupil to enter the eye is

$$m\theta \le \tan^{-1}\left(\frac{p}{\ell}\right).$$

In the limit that the size of the exit pupil is negligible, which is rather good in this case, the three formulas become equal. With the small angle approximation and the approximate value for ℓ they give a half-field of view of

$$\theta \le \frac{p}{m\ell} \approx \frac{p}{f_2(m-1)}.$$

In this limit the field of view is proportional to the diameter of the eye opening. However, there are rays of light leaving the telescope that don't enter the iris of the eye. You can see these by moving your eye around, up or down, right or left. This gives you access to a much wider field of view without any movement of the telescope. It is often stated that Galileo was not able to see the whole face of the Moon without resetting his telescope. This is probably not true. For a fixed position of the eye, the field of view might typically be only 14 or 15 minutes of arc, whereas the angular size of the Moon is about 30 minutes. However, for the replica telescope that I constructed, the field of view that can be seen by "cheating" the eye around is about 1.6 degrees. The telescopes that Galileo used for his discoveries have not survived. I can calculate the total accessible field of view for the two telescopes that are attributed to him. One has magnification of 20 but the eyepiece is generally believed to be a replacement. For that telescope I calculate a total field of just over one degree. For the other telescope, which was on display at the Franklin in Philadelphia in 2009, the magnification is 14 so that it does not seem to have been one that he used for his discoveries. It has quite a small eyepiece. I calculate the total field accessible without moving the telescope to be just over 30 minutes.

The total field of view that can be accessed is determined by the diameter of the eyepiece. Figure 25.16 is very similar to the second case that I worked through, except that the vertical line on the left is now the width of the eyepiece. I have used p' for the half-height of the eyepiece, and set its distance from the exit pupil equal to $f_2(1-1/m)$, which is exact in this case. There are, of course, formulas corresponding to the three limits I used earlier, but if I neglect the size of the exit pupil and make a small angle approximation they reduce to

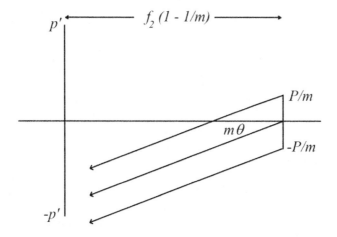

Figure 25.16 Field of view with eyepiece as field stop

$$\theta \leq \frac{p'}{f_2(m-1)} = \frac{p'}{f_1 - f_2}.$$

Actually, this is almost exactly the same as the formula for the field of view of a Keplerian telescope except that the separation of the lenses in that case is the sum of the focal lengths. If the eyepiece is very small, smaller than the opening of the observer's eye, this value of θ is smaller than the value containing the iris opening, p, and this is the limiting value even with the eye centered on the eyepiece.

Chapter 26: Four variable stars

Looking at variable stars is certainly not an original project. There is an organization, the American Association of Variable Star Observers (AAVSO) devoted to it. AAVSO produces invaluable charts and other literature to assist the observer. There is a training tutorial at www.aavso.org/10-star-training.

I was particularly interested in three stars, δ–Cephei, β-Lyrae, and β-Persei, also known as Algol. They all are linked to an interesting eighteenth-century astronomer named John Goodricke. He discovered the variability of δ-Cephei and β-Lyrae, and he made the first accurate determination of the period of Algol. There is a brief account of his life in "Sky and Telescope", volume 56 pp 400-403, November 1978. In a good sky all three of these stars are visible to the naked eye. There was a description of them and other naked-eye variables in an article by John Isles, in Sky and Telescope, May 1997, pp 80-82.

δ-Cephei is the prototype of the Cepheid variables that were used in an important method for determining the distances to stars and, in particular, to such variables in other galaxies. The key idea was the discovery by Henrietta Swan Leavitt that there was a one-to-one relationship between the periods of such stars and their absolute magnitudes. From a determination of the periods of variables observed in other galaxies, Edwin Hubble could deduce their absolute magnitudes and from a comparison of their absolute and apparent magnitudes he could find their distances. There is an account of those developments in "The Day We Found the Universe" by Marcia Bartusiak.

The other two stars belong to the category of eclipsing variables. The magnitude of β-Lyrae varies rhythmically as two stars rotate around and in front of each other. Algol is a more dramatic eclipsing variable star. Every three days a dark companion moves in front of Algol and it blinks out for a few hours. This odd behavior of Algol may have been known for perhaps a thousand years. Its Arabic name translates as "the ghoul" or "the demon star". However, its behavior was first explained by John Goodricke in 1782.

Over the years, I have looked at each of these stars several times, and actually watched Algol going through a minimum, but I had never tried to make a plot of the magnitudes. There are several ways to find the magnitude of a variable star. The oldest method is simple in principle. The AAVSO charts identify neighboring stars that are slightly brighter and slightly fainter than the variable, and give their brightnesses to the nearest tenth of a magnitude. By comparing the variable star with the two neighbors, it is possible to estimate the magnitude of the variable star. It is suggested in the AAVSO tutorial that it is quite easy to make estimates to two-tenths of a magnitude, and an experienced observer can estimate the brightness of the variable to the nearest tenth of a magnitude. More modern methods use some form of digital camera to record a star field around the variable star, and then a computer program to compare the magnitudes. I wanted to try the old-fashioned way first, partly because it was described by Lesley Peltier in his beautiful biography "Starlight Nights".

I have to say that it took me several tries to get the hang of the procedure. The two companion stars for δ-Cephei form a triangle with it, small enough that all three stars fit into the field of view of my binoculars. On the first evening, I decided that all three stars were equally bright! I got help from two books, "Astronomy with Binoculars" by James Muirden and "Observing Variable Stars" by David Levy. Muirden pointed out that I should not simply be comparing the three stars all in the same field of view, because I am using different parts of my retina to look at them. Rather, I should bring each of the stars to the center of the field of view for a few seconds and compare their brightnesses in that central position. Levy pointed out that in the case of δ-Cephei, the two comparison stars, ς- and ε-Cephei, are 0.8 magnitudes apart. I could simply put the variable in one of five categories: as bright as ς, a bit less bright than ς, about half-way between ς and ε, a bit brighter than ε, or about as bright as ε. These categories translate into steps of about 0.2 magnitudes and are a lot more intuitive to use. I used this simplified approach for a few nights, until I felt confident enough to start estimating the brightness to the nearest tenth of a magnitude. Another helpful tip was to defocus the binoculars slightly and compare the fuzzy circular images. I like to view both in and out of focus before making up my mind. Even if I could see the stars naked-eye in my Cherry Hill sky, I would still use binoculars because of this defocused view. Another problem I had was that I tended to be influenced by what the last reading had been. In the case of δ-Cephei, I made observations at about the same time every evening, to the extent that the weather allowed it. I found it difficult not to extrapolate mentally the values from the previous days and second-guess myself as to what the next reading would be. Actually, the New Jersey weather was helpful in this regard, because it was quite common for there to be breaks of several days between observations. After a two or three day break I could make an estimate with less pre-judgment.

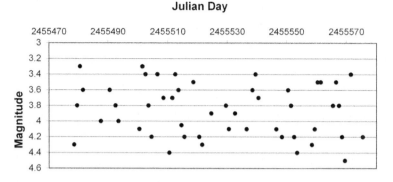

Figure 26.1 Raw data for δ-Cephei

I observed δ-Cephei over a three-month period in the Fall and Winter of 2010-2011, as long as it was fairly high in the northern sky. I plotted the values on a graph in Excel, but it wasn't very informative. There were many gaps, and also, the readings taken at one-day intervals didn't sample the regions of maximum or minimum intensity closely enough to show the shape of the light curve. What I needed was to reduce all the observations to one single period, so that the light curve would be sampled much more closely. This is rather easily done in a spreadsheet by using the "modulo", or MOD(,) function. I used the same idea when I was

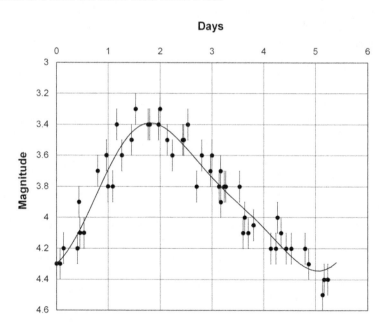

Figure 26.2 Data for δ-Cephei on reduced time scale

195

analyzing the moons of Jupiter. I had to put in a value for the period, and I used the handbook value of 5.36667 days. This may be cheating, but I could extract from my own data a period of 5.3 to 5.4 days, and I decided there was no point in not using the more accurate value. The results are shown in Figure 26.2. I have plotted error bars with the uncertainty taken as 0.1 magnitudes. This is clearly about right, based on the scatter of the points. I also had the computer draw a line through the curve. What I actually did was to make up a functional form that I thought would do the job. I used a five-term Fourier series. The five constants in the expression were chosen to give the best overall fit, using standard statistical procedures. The resulting line is my best estimate of the light curve. It is close enough to the published curves that I feel I now have a new friend up in the sky.

δ-Cephei is a genuinely variable star. It is a red super-giant that undergoes massive pulsations, on a time scale of a few days, as its intensity varies. I found a more complete description in "An Introduction to Modern Astrophysics" by Bradley W. Carroll and Dale A. Ostlie. Some variable stars work differently; they are called eclipsing variables. In them, a bright star and a fainter star are orbiting around each other. For us to see the effect they need to be orbiting in a plane that is more or less edge on seen from the Earth. When the fainter star moves in front of the brighter one, we see a drop in brightness, and there can be a second drop when the fainter star is obscured. β-Lyrae and Algol are both eclipsing variables but with quite different light curves.

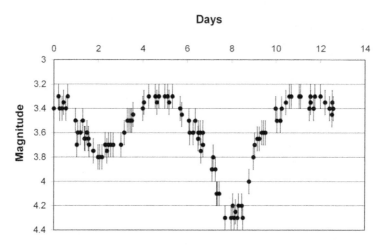

Figure 26.3 Data for β-Lyrae plotted on a reduced time scale

I watched β-Lyrae over an eight-month period in the summer and fall of 2011. Figure 26.3 shows the results, already plotted using the MOD(,) function to reduce the data to a single period. I again used the handbook value, of 12.4 days, although I had a pretty good estimate of it from my own observations. I was getting better at making the brightness estimates by this time. The scatter of the points, both for β–Lyrae and Algol, is less than for δ-Cephei. The light curve this time is totally different from that of δ-Cephei. In particular, there are two minima in each cycle. Each minimum corresponds to one star blocking the light from the other. At the maxima, we see the light from both stars. There is a flat region at each maximum because the two stars do not overlap for some time. In fact, the flat regions are quite narrow for β-Lyrae, and this indicates that the distance between the two stars is only slightly larger than the sum of their radii. Also, the time separation of the two minima is not very different from half a period and this indicates that the two stars are of similar size. Obviously, there is a lot of information to be pulled out of data like this, but it takes quite a lot of analysis and it really needs better quality data.

I followed Algol, β-Persei, over the same period that I was watching β-Lyrae. In fact I could keep taking observations through the winter of 2011-2012. Here is the light curve, plotted using the handbook value of 2.8 days for the period.

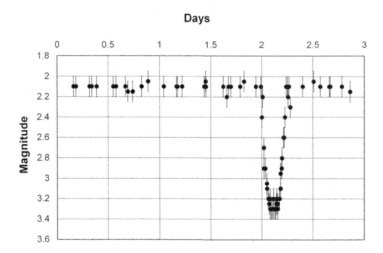

Figure 26.4 Data for Algol plotted on a reduced time scale

This clearly differs from the light curve for β–Lyrae in that for most of the period the intensity is more or less constant. This tells you that the two stars that form the double are a long way apart compared with their sizes. Most of the time, you see the light coming from both of the stars. Also, my results do not show clearly the double minimum structure of the light curve. If I compare my results with curves available on the internet, the second minimum is a very weak one at a time around 0.7 days, and I can just convince myself, though perhaps not you, that I have seen it. I also hoped to measure clearly the shape of the main minimum, whether it has a round or a flat bottom. This would tell me something about the relative sizes of the two stars. To measure a complete minimum, you have to wait until it all takes place during the nighttime hours at your site. I observed two minima fairly thoroughly, one completely and one about three-quarters complete. In one case my points showed a flat bottom and in the other case a rounded, or even pointed minimum, but, when I put in the tenth-of-a-magnitude error bars, I saw that either line would go through both sets of points. I have reached the limit of the estimation method.

I mentioned that it is also possible to use technologically more advanced methods. The next step up is to use star field photographs taken with a DSLR. The technique was described briefly in an article in "Sky and Telescope", April 2011 issue, pp 64-67. More details are given in the book by Robert K. Buchheim, listed in the references, and in a recent article in the December 2020 issue of "Sky and Telescope", pp 60-65. If I want to extend my measurements on Algol, I shall move on to one of these methods. But I am happy that I spent some time with the old-fashioned approach. It really does give you a feeling of being familiar with the stars.

The fourth star that I followed is perhaps the grandfather of all variable stars, Mira. Its name implies "miracle". Figure 26.5 shows two photographs of a small part of the constellation Cetus.

Figure 26.5 Part of the constellation Cetus

The left-hand photograph was taken on November 19th, 2017 and the right-hand picture was taken five weeks later on December 26th. Mira is the bright star in the right-hand picture. By comparing it with other stars outside this small field of view, I estimated it to be just about third magnitude in brightness. On the original photographs it is just visible in the left-hand picture and I estimated it to be of ninth magnitude. To a naked-eye observer the new star would have appeared from nowhere. No wonder it was called miraculous Unfortunately, in my backyard sky it is still not naked-eye visible even at its brightest, so I followed it through its maximum by taking photographs and comparing Mira with other stars in the field. I judged that I could estimate its brightness to about half a magnitude. This is a lower precision than I claimed for my earlier measurements but the range of brightnesses involved is also much larger. Mira changes brightness by about six magnitudes, or a factor of 250.

I knew that the period of Mira was about eleven months. It is conveniently placed in my sky from November to May, so that each appearance I saw a later part of the curve. The next figure shows three consecutive appearances.

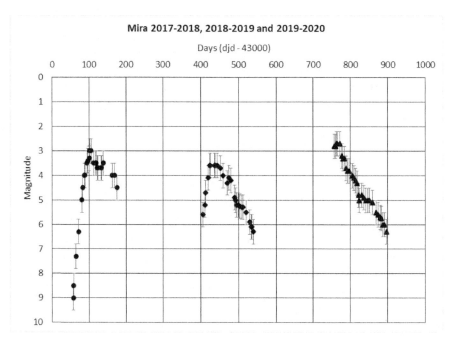

Figure 26.6 Three maxima of Mira

Mainly Optics

I can make a guess at the period and slide the middle curve to the left by this amount and the right-hand curve by twice the guess. A value of 331 days works best.

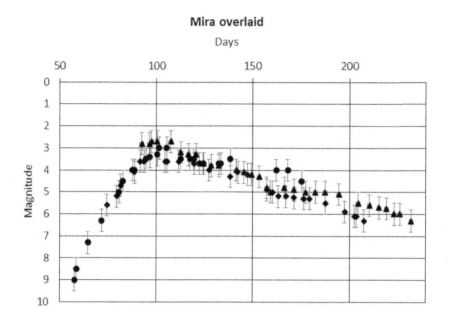

Figure 26.7 Maxima shifted by 331 and 662 days

There are many more variable stars for me still to follow!

Appendices

A: Trigonometry formulas

The commonly met trigonometric functions can be defined in terms of a right-angled triangle. The side opposite the right angle is the hypotenuse, labeled r. The other two sides can be referred to by their relation to the angle labeled θ. y is the side *opposite* to θ, x is the side next to, or *adjacent* to, θ. The lengths of the three sides satisfy Pythagoras's theorem

$$r^2 = x^2 + y^2.$$

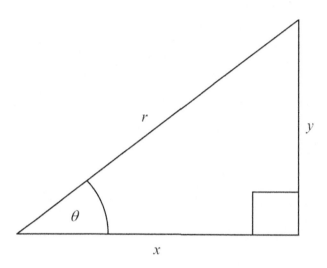

Figure A.1 Right-angled triangle

The sine, cosine, and tangent of θ are defined as the ratios of sides

$$\sin(\theta) = \frac{y}{r},$$

$$\cos(\theta) = \frac{x}{r},$$

$$\tan(\theta) = \frac{y}{x}.$$

This picture doesn't make sense if θ is bigger than 90°. The sum of the angles of a triangle is 180° and the right angle already uses up 90° of that. In fact, it is possible to extend the definitions of the functions to apply to angles greater than 90° and to negative angles. I need them in this way when I make extended plots of oscillating quantities such as the angular size of the Moon.

Appendices

The simplest way to handle this is to resort to a scientific calculator or spreadsheet that has the functions built in!

A formal solution to, for example, the last equation is

$$\theta = \tan^{-1}\left(\frac{y}{x}\right).$$

In other uses, the superscript -1 means "one over", but in this trigonometric usage the right side of the equation can be read as "the angle whose tangent is y/x". Again, scientific calculators have the inverse tangent and other inverse functions built into them.

The most familiar unit for angles is the degree. There are 360° in a full rotation. If necessary, a degree can be divided into 60 minutes and a minute into 60 seconds. This is confusing notation since it sounds like the units for time. One way to remove the ambiguity is to write the angular units as "minutes of arc" and "seconds of arc", or "arc-minutes" and "arc-seconds". It just sounds a bit pedantic.

In many mathematical and scientific applications a different unit for angles is used, called the *radian*. The radian is defined in terms of the arc length s in this picture

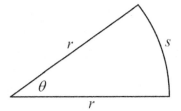

Figure A.2 Radians defined

The radius of the circle is r and s is the distance measured round the curved arc, or the arc length. The angle θ in radians is defined as

$$\theta = \frac{s}{r} \text{ radians.}$$

This might be a good place to remind you of a set of formulas, involving the number π, which crop up many times in spherical geometry:

202

the circumference of a circle $= 2\pi r$,
the area of a circle $= \pi r^2$,
the surface area of a sphere $= 4\pi r^2$,
the volume of a sphere $= 4\pi r^3/3$.

Now, in the picture above, imagine extending the arc to form a complete circle. The arc length would then be the circumference of the circle, which is 2π times the radius. But a complete circle corresponds to an angle of 360° at its center, so the conversion between degrees and radians is

$$360^o = \frac{2\pi r}{r} = 2\pi \text{ radians.}$$

To convert an angle from radians to degrees, multiply it by 360°/2π and to convert an angle in degrees to radians multiply it by 2π/360°. 1 radian is approximately 57.3°.

In many astronomical applications, the triangles involved turn out to be very skinny.

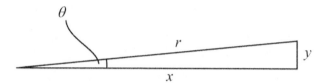

Figure A.3 A skinny triangle

If the opposite side to the angle θ is small, then the sine and tangent of θ are also small, and the length x is not much smaller than r. The values of all the trigonometric functions can be approximated

$$\sin(\theta) \approx \tan(\theta) \approx \theta \text{ in radians,}$$
$$\cos(\theta) \approx 1.$$

Another way of thinking about this is that for such a skinny triangle, it doesn't make much difference if the side y is replaced by a short arc of a circle. In several of the math parts of this book, I use the tangent function and its small angle approximation quite a lot. For example, in chapter 19 I measure the average angular size of the Sun as 31' 55", or 0.00928 radians. I then approximate the diameter of the Sun as this angular size multiplied by its distance, which is just 1 in astronomical units, so that the Sun's diameter is 0.00928 a.u. For such small angles, the error introduced by this approximation is very tiny compared with the uncertainty in my measurement.

Appendices

Another use of trig functions is to find a *component* of a displacement or distance. In chapter 9 I needed the north-south distance from Cherry Hill to West Cape May and I had measured the mileages for three legs of the drive.

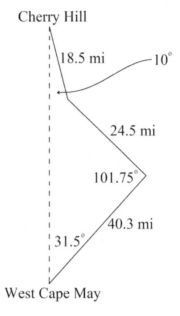

Cherry Hill

18.5 mi   —10°

24.5 mi

101.75°

40.3 mi

31.5°

West Cape May

Figure A.4 Vector example – same as figure 9.1

Remember that the side of a right-angled triangle adjacent to the angle θ has a length equal to the hypotenuse times the cosine of θ. So, if you picture a right angled triangle at the top of this diagram, with a hypotenuse of 18.5 miles, the vertical side is $18.5 \cos(10°) = 18.2$ miles. In a similar way, a right angled triangle at the bottom of the picture, with a hypotenuse of 40.3 miles, has a vertical side of $40.3 \cos(31.5°) = 34.4$ miles. The middle part needs an extra step, because I need to know the angle between the sloping line with length 24.5 miles and the vertical. The sloping line makes an angle of 101.75° with the 40.3 mile leg, and the angles of a triangle add up to 180°. So the angle I need is $180 - 101.75 - 31.5 = 46.75°$, and the vertical component of that leg is $24.5 \cos(46.75) = 16.8$ miles. The total vertical distance is $18.2 + 34.4 + 16.8 = 69.4$ miles.

Appendix B: Thin lens formulas

If parallel rays of light are shone into a thin lens from the left side they are refracted in the lens and come together at a focal point F on the right side. The distance from the lens to the point F is called the focal length f of the lens.

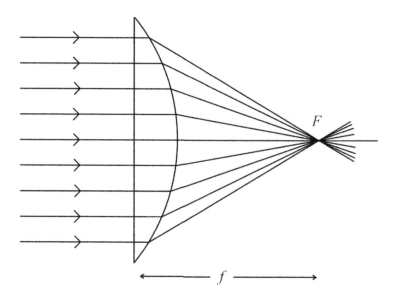

Figure B.1 Focal point of a converging lens

If the light is shone into the lens from the right the rays come to a focus at a second focal point on the left side of the lens, shown in Figure B.2. Interestingly, and a bit surprisingly, the two focal points are the same distance from the lens, even if the lens is not symmetrical. Both of the statements I have just made are only approximately correct. The rays come to a sharp focus if the focal length of the lens is much greater than its diameter. If the focal length is comparable to the diameter, or even less that about ten times the diameter, so-called spherical aberration blurs the sharpness of the focus. Also, the two focal points are the same distance from the lens only to the extent that the thickness of the lens is small compared with the focal length.

Now suppose that an object is placed a finite distance from the lens, but outside the focal point on that side. The height of the object is h_o. The lens forms an image of the object on the other side of the lens. The height of the image is h_i and you can see from the Figure B.3 that it is upside down.

Appendices

There are two very powerful formulas that relate the positions and heights of the object and the image.

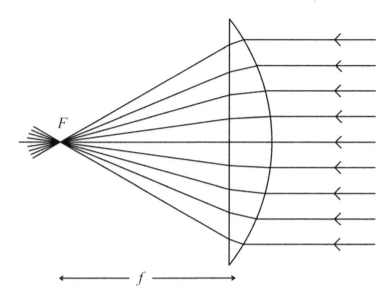

Figure B.2 Second focal point of a positive lens

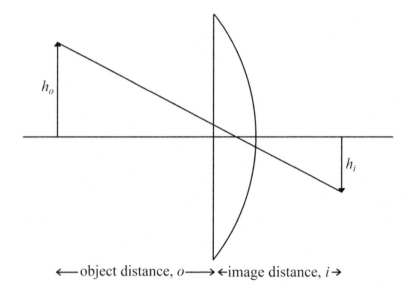

Figure B.3 Image location for a positive lens

206

The distance, o, of the object from the lens and the distance, i, of the image from the lens are related by the formula

$$\frac{1}{o} + \frac{1}{i} = \frac{1}{f},$$

and the heights of the image and the object are related by

$$\frac{h_i}{h_o} = (-)\frac{i}{o}.$$

A minus sign is usually put into this equation because the image is upside down compared with the object. The idea is to convey as much information as possible. If these formulas are new to you, it is worth playing with them. Put in some numbers. If you choose the focal length to be 12 cm and the object distance to be 36 cm, you will find the image distance to be 18 cm. If the object distance is changed to 24 cm, the image distance will become 24 cm also. If the object distance is made very large, say 120 cm, the image distance is approximately 13.3 cm. In fact, as the object distance get larger and larger, the image distance gets closer and closer to the focal length. In an astronomical telescope, the object is so far away from the objective of the telescope that the image is formed right at the focal point.

If you set the object distance to be less than the focal length, a new idea appears. If the object distance is 4 cm, the image distance is calculated to be -6 cm. Can a negative value for the image distance have a meaning? It turns out that the image is actually located *behind* the lens; that is, it is on the same side of the lens as the object. Furthermore, the rays of light don't go through the image but only seem to spread out from it when seen from the other side of the lens. This is called a *virtual* image. The understanding is, then, that a negative distance indicates a virtual point. The same idea can be applied to focal points and even, in multiple lens systems, to an object. The negative value for i cancels the minus sign in the formula for the height of the image. The image is the right way up, or erect. When you use a positive lens as a magnifying glass you are creating a virtual image; the image you are looking at is on the same side of the lens as the object being magnified. A telescope eyepiece works in the same way.

The lens I first drew is a converging lens. It is thicker in the middle than at the edges. A lens that is thinner in the middle is a diverging lens. If parallel rays of light enter a diverging lens from the left, they spread out on the right side of the lens. Amazingly, you can use exactly the same formulas as for the converging lens with the simple expedient of assigning a negative focal length to the diverging lens. Of course, the defining characteristic of a Galilean telescope is that it uses a negative or diverging lens for the eyepiece.

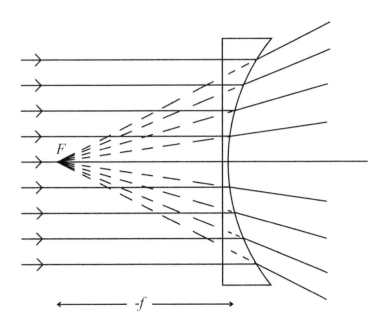

Figure B.4 Focal point of a diverging lens

The same formulas can be applied to curved mirrors. They also come in two flavors. Concave mirrors are assigned a positive focal length. They are used as the objectives in Newtonian telescopes and in the more complex Schmidt-Cassegrain and Maksutov-Cassegrain telescopes. Convex mirrors behave in some respects like diverging lenses, and are assigned negative focal lengths.

As a final comment on the power of these equations, the lenses we use for the objectives and eyepieces of modern telescopes are combinations of several lenses close together. It is possible to rescue the two basic formulas by defining two planes, the *principal planes*, somewhere in the combination. Distances from the front of the lens system are made to the front principal plane and distances behind the lens are made to the rear principal plane. The two simple formulas then apply. An interesting, and extreme, application of this idea in photography is the telephoto lens. This has a very asymmetric design such that the rear principal plane is much further away from the film or sensor than is the actual back of the glass elements. This is explained in more detail in, for example, "Optics", by Eugene Hecht, listed in the bibliography.

In an optical system that consists of a series of separate lenses or mirrors, one way to attack the analysis is to find the image that would be created by the first lens alone, then to use this as the object for the second lens, and so on through the system. It is in this type of analysis that the situation comes up that the image created at some stage is not actually formed. The next lens in the series is too close. The procedure is to calculate where the image would have been formed, and to use this as the virtual object in the second stage. Again, Hecht's book gives examples.

Appendix C: Error Analysis

I have tried to include some estimate of the uncertainty in my measured values, and I felt that I needed to add a short appendix on the calculation and manipulation of the uncertainties. A common terminology is to use the word "errors", but I much prefer "uncertainties", which seems to convey better the idea that there is a spread of values, or a range, in which the correct answer probably lies. There are some more or less complicated formulas for finding error estimates and for combining them. Unless you actually enjoy doing it, I am not trying to push you into that. We are doing this because it is fun, remember!

Uncertainties can arise from two sets of causes; there are random errors and systematic errors. An example of a random error is the uncertainty associated with a stop-watch timing. Stop-watches are usually very accurately calibrated, but if you repeat a timing measurement you will generally not get exactly the same value every time. There is a small human error involved in starting and stopping the stop-watch. Random errors are described well by the theory of probabilities and all the rules I shall give for manipulating errors are based on this.

The second type of error is associated with some systematic fault in the procedure. The measuring device might not be properly calibrated. If you have ignored my advice and accepted the nominal value for the focal length of a lens, you probably have a systematic error in those numbers. If you use a ruler to measure a length, you rely on the ruler being properly calibrated. A plausible guess at the uncertainty of such a measurement would be the smallest division on the scale. If an inexpensive ruler is marked out in centimeters and millimeters, probably ± 1 mm is the best you can hope for. Try putting several rulers side by side and see how well they agree. A systematic error might also creep in if you use an incorrect or approximate theory to analyze the data.

Suppose that you have made a measurement that can be repeated several times and you have found a spread of values. If the variations are randomly distributed (and in the absence of any other clues you have to assume that they are) the best estimate of the true value is the average of the readings. Add them together and divide by the number of them to get the average, also called the mean value. If you have only a few readings, two or three or four, the safest estimate of the uncertainty is the difference between the average and the value most different from it. Suppose I have dreamed up some way of measuring π and I get a set of results: 3.1, 3.4, 3.3. The average of these is 3.267. The first value deviates from this by 0.167, and this is the

Appendices

largest deviation. I might say that the best answer is 3.267 ± 0.167, but most scientists would argue that the decimals beyond the first figure have no meaning, and would round the result to 3.3 ± 0.2.

If you are in a position to repeat a measurement many times, say ten or twenty times, then you can use more elaborate formulas. Suppose you have n measurements of a quantity x. Label them with an index i, as x_i, with the value of i running from 1 to n. The best estimate of the true value is still the average, equal to the sum of the x_i divided by n. This is represented by x with a bar over it. As an equation, this is written

$$\bar{x} = \frac{1}{n}\Sigma x_i.$$

The upper case Greek Σ (sigma) means "sum all values of".

An estimate of the spread of values, called the standard deviation and represented by a lower case sigma, σ, is found by taking the differences between each of the observations and the mean. Some of the differences are positive and some are negative. Square each of them, to get a positive number, sum the squares and take the average. The actual formula is

$$\sigma^2 = \frac{1}{n-1}\Sigma\left(x_i - \bar{x}\right)^2.$$

The formula has $(n-1)$ rather than just n to allow for the fact that we have only an estimate of the true central value rather than the limit that we would get with a very large number of observations. This is an estimate of the spread, but the value of the mean is defined rather better than this. The standard error of mean is smaller by a factor of the square root of the number of observations.

$$\sigma_m = \frac{\sigma}{\sqrt{n}}.$$

Quite an important idea is that the true answer probably lies in the range $\pm\sigma_m$ but it might lie outside it. If the errors are truly random, there is about a 15% chance that the true answer is above this range and a 15% chance that it is below it and therefore a 70% chance that it lies within the range. Some cautious experimenters quote $2\sigma_m$ uncertainties so that they can feel more nearly confident that they are spanning the correct value.

Once you have uncertainties for the individual quantities, there are three rules to follow when you combine them in formulas.

1) When two experimental values are added to or subtracted from each other, square their uncertainties, add the squares, and take the square root.

2) When two experimental values are multiplied or divided by each other, first find their percentage or fractional uncertainties, square them, add them together, and take the square root. This is the percentage or fractional uncertainty of the final answer.

3) When a number is raised to a power (such as squared or cubed) the percentage uncertainty is multiplied by the power. This also works for fractional powers. If you take the square root of a number, this corresponds to a power of one half, and the percentage uncertainty is halved.

You will find that in cases where one uncertainty is much larger than the other the larger uncertainty dominates the answer. For example, if there are two values, 100 ± 1 and 40 ± 10, the rule gives 140 for the sum, and the uncertainty is

$$\sqrt{1^2 + 10^2} = \sqrt{101} = 10.05 \approx 10.$$

If you want to read a fuller description of these ideas, a very good book is "An Introduction to Error Analysis" by John R. Taylor. It is written for a first-year college science student with no background in measurement but with a reasonable high-school math background. Most of the results I use are in the first four chapters, which use only algebra. But, as I said, you are doing this for fun.

Appendix D: Fitting formulas and parameters

In many of the cases I have written about, the end value came from fitting a formula to the measurements. The simplest situation is that the measured values are expected to lie on a straight line and you need the value of the slope of the line. For example, in the case of the synodic and sidereal moon measurements, I had a series of dates and I assumed that the dates went up linearly with the number of months. The standard criterion for finding the best fitting line is called the principle of least squares. You typically will have a series of observations and an equation, containing one or more constants, which you can use to calculate values that should match the observations. Some of the calculated values will be larger than the observed ones and some will be smaller. Also, the uncertainties of the various observations may differ. A measure of how well the equation reproduces the measurements is the sum

$$\chi^2 = \sum \left(\frac{\text{measured value - calculated value}}{\text{uncertainty}} \right)^2.$$

The sum is over all the observations. This type of sum is usually called χ^2 (chi–squared) in statistical analysis. The method of least squares is to vary the constants in the equation until χ^2 is as small as possible.

In the commonest situation, of a sloping straight line, the equation is of the form

$$y = A + Bx.$$

The least-squares condition for the constants A and B can be solved exactly. The formulas are given, for example, in the book by John R. Taylor that I recommended in the previous appendix. If you are working with Excel, it, happily, knows all about these formulas. On the chart options you can select to add a trend-line, which can be linear or a polynomial, and you can choose to display the equation on the chart. There are also formulas available for the uncertainties in A and B. Excel also knows about these, but you have to use the array function LINEST(). Before I knew that such a function existed I got in the habit of programming the necessary bit of the spreadsheet myself and simply copying it from one application to another. The formulas are again in Taylor's book.

The method is easily extended to the case of a polynomial such as

$$y = A + Bx + Cx^2 + \ldots$$

215

Appendices

The key feature of this is that, while y is no longer a linear function of x, it does depend linearly on each of the constants, A, B, C etc. This is what allows the minimization condition to be solved exactly. Again, Excel knows all about this.

The situation is more complicated when the calculated quantity depends non-linearly on the constants. For example, in the analysis of Jupiter's moons, I used a formula

$$y = d \sin\left(\frac{2\pi t}{T} + \varphi\right).$$

The distance y of the moon from Jupiter does depend linearly on the radius of its orbit, d, but it clearly depends in a more complicated way on the two parameters T and φ. The procedure now is to start from as good an estimate of the parameter values as you can get, so that the corrections to those values are expected to be small. A basic idea from calculus is that, if the corrections are small, the calculated quantity will be approximately linear in the changes. So you can set up equations very similar to those for a multi-parameter linear fit, but the equations are now to calculate just the *changes* in the parameters. The catch is that the linearity is only approximate so that the changes calculated are only approximate also. They should give better values than the original estimates, so the procedure is to repeat, or iterate, the calculation several times and the values should finally converge to the set giving the minimum value of χ^2. There is an entry in Wikipedia describing the method, under the heading "Least Squares", but the clearest account I have read is the one I first learned it from, in a rather old book, "The Mathematics of Physics and Chemistry" by Henry Margenau and George M. Murphy. Secondhand copies of this are still available. Actually, the book is slightly younger than I am, so I won't call it *very* old!

To illustrate the method, let me apply it in more detail to the fit to the sine function case. I actually used this in several parts of this book. To start the fit, you need estimates of the three parameters. Call these d_0, T_0, and φ_0. The closer these are to the best values, the faster the process will converge. I found them by using the trick of plotting the measured points in Excel, with the times of the measurements reduced to a single period by means of the MOD(,) function. Simply put different guesses for the period into the cell assigned to its value, until the curve looks as smooth as possible. The least-squares method finds corrections to these initial values. Call these δd, δT, and $\delta\varphi$. If these corrections are small, the changes in the calculated values of y are linear in the parameter changes. That is

$$y = y_0 + A\delta d + B\delta T + C\delta\varphi,$$

where y_0 is the value of y calculated with the initial parameters, and A, B, and C are the derivatives of y with respect to the three parameters, also evaluated with the initial parameter values.

$$A = \sin\left(\frac{2\pi t}{T_0} + \varphi_0\right),$$

$$B = -\frac{2\pi d_0}{T_0^2}\cos\left(\frac{2\pi t}{T_0} + \varphi_0\right),$$

$$C = d_0 \cos\left(\frac{2\pi t}{T_0} + \varphi_0\right).$$

The key idea is that the equation for y is linear in the δ's, so that the solution is essentially the same as for the linear fit problem. The catch is that the coefficients A, B, and C are functions of the parameter values, and therefore they have changed. The procedure is to do the calculation again, and to keep repeating it until the values stop changing. It may take a dozen or so iterations, so it is worth setting it up in a medium-sized computer program. I have a feeling that it should be possible to set it up as a spreadsheet, but I haven't tried to do it.

I have used this method in the past to fit twenty or thirty parameters to an extensive set of data, and, for a problem of that complexity, you do need the full procedure that I just outlined since it adjusts all the parameters simultaneously. The applications in this book involved only two to four parameters. For some of these I used a much cruder approach. I wrote a routine to calculate χ^2 for a given set of parameters, and surrounded this with a simple loop that stepped through a range of values of each parameter in turn, where I manually selected the value that gave the lowest value. I still had to iterate the process several times, but it was not at all arduous. I don't see why you shouldn't be able to do that version in a spreadsheet also, although I haven't attempted that.

Appendix E: The line goes through the points

Here are three quotations from far greater minds than mine:

"Mathematics is the gate and key of the sciences. …Neglect of mathematics works injury to all knowledge, since he who is ignorant of it cannot know the other sciences or the things of this world."
Roger Bacon – Opus Majus, ca 1265, quoted in Wikipedia.

"Philosophy is written in this grand book, the universe, which stands continually open to our gaze. But the book cannot be understood unless one first learns to comprehend the language and read the letters in which it is composed. It is written in the language of mathematics, and its characters are triangles, circles, and other geometric figures without which it is humanly impossible to understand a single word of it; without these, one wanders about in a dark labyrinth."
Galileo Galilei, "The Assayer" (Il Saggiatore), 1623. Translated by Stillman Drake in "Discoveries and Opinions of Galileo", Doubleday, 1957.

"The fact that mathematics does such a good job of describing the Universe is a mystery that we don't understand. And a debt that we will probably never be able to repay."
Lord Kelvin

Notice the date of the first quotation, 1265. I find that amazing.

The third quotation concerns the idea that most scientists believe that they are measuring the values of natural phenomena, which are not influenced by the act of measurement (except for small quantum mechanical effects), whereas the mathematics used in the theoretical analysis is a product of the human mind rather than something natural that we have discovered. But it often seems that the two fit each other uncannily well. I have my own version of this philosophical idea, and you can verify it for yourself. Set up a pendulum consisting of a small weight on the end of a string, and measure its period, the time it takes the pendulum to swing backwards and forwards through a complete cycle. Actually, you should measure the time for a large number of swings, and divide the time by the number to get the time for a single swing. Repeat the measurement for several different lengths of the string. A graph of the period as a function of the length of the string is plotted in Figure E.1. This is not a waste of time. I used these measurements to extract one of the ingredients in my determination of the mass of the Earth.

219

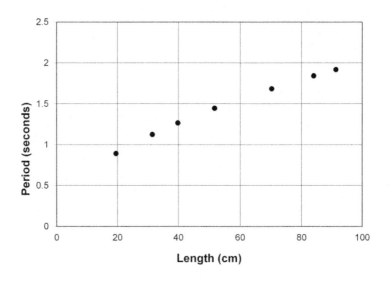

Figure E.1 Measured period of a simple pendulum

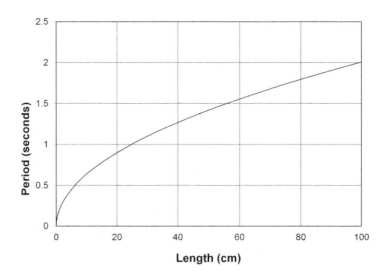

Figure E.2 Theoretical period of a simple pendulum

The simple pendulum is a standard example used in all introductory physics text books. If the amplitude of the swing is not too big, the period T is given theoretically by

$$T = \frac{1}{2\pi}\sqrt{\frac{L}{g}},$$

where L is the length of the pendulum and g is a physical constant, the acceleration due to gravity, approximately equal to 9.8 m/s^2. The value of g varies slightly from one place on Earth to another. I analyzed my results to get a value of 9.80 m/s^2 at my home. I can plot a graph of this as well. It is plotted in Figure E.2.

The magic trick is to plot the experimental points, *which are a measurement of nature*, and the mathematical line, *which is a product of our analysis*, on the same graph:

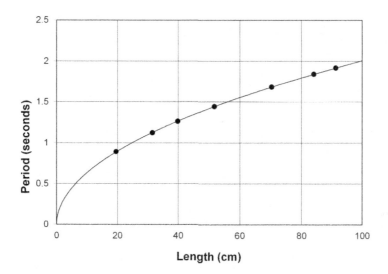

Figure E.3 And the line goes through the points!

Bibliography

Bartusiak, Marcia, "Archives of the Universe", Pantheon Books, New York, 2004.

Bartusiak, Marcia, "The Day We Found the Universe", Pantheon Books, New York, 2009.

Brewer, Sydney G., "Do-it-yourself Astronomy", Edinburgh University Press, 1988

Buchheim, Robert K., "Astronomical Discoveries You Can Make, Too!", Springer, 2015.

Carroll, Bradley W. and Ostlie, Dale A., "An Introduction to Modern Astrophysics, Addison-Wesley Publishing Company Inc., 1996.

Clark, John D., "Measure Solar System Objects and Their Movements for Yourself", Springer, 2009.

Copernicus, Nicolaus, "On the Revolutions of Heavenly Spheres", reprinted in "On the Shoulders of Giants", ed. Stephen Hawking, Running Press, Philadelphia and London, 2002.

Cowley, E. Roger, "A classroom exercise to determine the Earth-Moon distance", American Journal of Physics, volume 57, page 351, April 1989.

Dreyer, J.L.E., "A History of Astronomy from Thales to Kepler", Dover Publications, 1906, 1953.

Duffett-Smith, Peter, "Astronomy with your personal computer", Cambridge University Press, 1986.

Duffett-Smith, Peter and Zwart, Jonathan, "PracticalAstronomy with your Calculator or Spreadsheet", 4th edition, Cambridge University Press, 2011.

Galilei, Galileo, "The Starry Messenger", translation published in "Discoveries and Opinions of Galileo" by Stillman Drake, Doubleday Anchor Books, 1957.

Hecht, Eugene, "Optics", 4th edition, Addison-Wesley, 2002.

Bibliography

Hughes, H. A. and Everitt, P. F., "On the Field of View of a Galilean Telescope", Trans. Opt. Soc. 22, 15-19 (1920).

Karkoschka, Erich, "Earth's Swollen Shadow", Sky and Telescope, pp 98-100, September 1996.

King, Henry C., "The History of the Telescope", Dover reprint ISBN 0-486-43265-3.

Levy, David, "David Levy's Guide to Variable Stars", 2nd edition, Cambridge University Press, 2005.

Margenau, Henry and Murphy, George M., "The Mathematics of Physics and Chemistry", 2nd edition, Van Nosrand Publishing Company, 1956.

Meeus, Jean, "Astronomical Algorithms", 2nd edition, Willmann-Bell Inc., 2009.

Meeus, Jean, "Astronomical Tables of the Sun, Moon and Planets, 2nd edition, Willmann-Bell Inc., 1997.

Meeus, Jean, "Mathematical Astronomical Morsels", Willmann-Bell Inc., 1997.

MICA - "U.S. Naval Observatory Multiyear Interactive Computer Almanac", Willmann-Bell Inc., 2005.

Muirden, James, "Astronomy with Binoculars", Arco Publishing Inc., New York, 1984.

Newton, Isaac, "(The) Mathematical Principles of Natural Philosophy", reprinted in "On the Shoulders of Giants", ed. Stephen Hawking, Running Press, Philadelphia and London, 2002.

Newton, Isaac, "Opticks", originally published 1730, Prometheus Books, 2003.

"Observer's Handbook", published annually by the Royal Astronomical Society of Canada.

Pannekoek, A., "A History of Astronomy", Dover Publications 1961, 1989.

Peltier, Leslie C., "Starlight Nights", Sky Publishing Corporation 1965.

Reeves, Eileen, "Galileo's Glassworks". Harvard University Press, 2008.

Renner, Eric, "Pinhole Photography – Rediscovering a Historic Technique", Focal Press, 1995.

Shull, Jim, "The Hole Thing", Morgan and Morgan Inc., 1974.

Sobel, Dava, "Galileo's Daughter", Walker and Company, New York, 1999.

Taylor, John R., "Introduction to Error Analysis", 2nd edition, 1997, University Science Books.

Tirion, Will and Sinnott, Roger W., "SkyAtlas 2000.0", 2nd edition, 1998, Sky Publishing Corporation.

"(The) Universe at Your Fingertips" Project Astro, Astronomical Society of the Pacific, 1995.

Waxman, Jerry, "A Workbook for Astronomy", 1984, Cambridge University Press.

Index

Index

Index

Printed in Great Britain
by Amazon